Winds

ADAM ALTER

Adam Alter is an Associate Professor at New York
University's Stern School of Business, and holds
an affiliated appointment within the Psychology
Department. He is the author of the *New York
Times* bestseller *Drunk Tank Pink: And Other
Unexpected Forces That Shape How We Think,
Feel and Behave* and has written for the *New
York Times*, *New Yorker*, *Atlantic*, *Wired*, *Popular
Science* and numerous other publications.

ALSO BY ADAM ALTER

Drunk Tank Pink

ADAM ALTER

Irresistible

Why You Are Addicted to Technology
and How to Set Yourself Free

VINTAGE

1 3 5 7 9 10 8 6 4 2

Vintage
20 Vauxhall Bridge Road,
London SW1V 2SA

Vintage is part of the Penguin Random House group of companies
whose addresses can be found at global.penguinrandomhouse.com.

Penguin
Random House
UK

First published in Vintage in 2017
First published in hardback by The Bodley Head in 2017

penguin.co.uk/vintage

A CIP catalogue record for this book is available from the British Library

ISBN 9781784701659

Printed and bound by Clays Ltd, St Ives Plc

Penguin Random House is committed to a sustainable future
for our business, our readers and our planet. This book is made
from Forest Stewardship Council® certified paper.

MIX
Paper from
responsible sources
FSC
www.fsc.org
FSC® C018179

For Sara and Sam

CONTENTS

PART 3

THE FUTURE OF BEHAVIORAL ADDICTION
(AND SOME SOLUTIONS)

IRRESISTIBLE

Prologue:

Never Get High on Your Own Supply

A t an Apple event in January 2010, Steve Jobs unveiled the iPad:

> What this device does is extraordinary . . . It offers the best way to browse the web; way better than a laptop and way better than a smartphone . . . It's an incredible experience . . . It's phenomenal for mail; it's a dream to type on.

For ninety minutes, Jobs explained why the iPad was the best way to look at photos, listen to music, take classes on iTunes U, browse Facebook, play games, and navigate thousands of apps. He believed everyone should own an iPad.

But he refused to let his kids use the device.

n late 2010, Jobs told *New York Times* journalist Nick Bilton that his children had never used the iPad. "We limit how much technology our kids use in the home." Bilton discovered that other tech giants imposed similar restrictions. Chris Anderson, the former editor of *Wired*, enforced strict time limits on every device in his home, "because we have seen the dangers of technology firsthand." His five children were never allowed to use screens in their bedrooms. Evan Williams, a founder of Blogger, Twitter, and Medium, bought hundreds of books for his two young sons, but refused to give them an iPad. And Lesley Gold, the founder of an analytics company, imposed a strict no-screen-time-during-the-week rule on her kids. She softened her stance only when they needed computers for schoolwork. Walter Isaacson, who ate dinner with the Jobs family while researching his biography of Steve Jobs, told Bilton that, "No one ever pulled out an iPad or computer. The kids did not seem addicted at all to devices." It seemed as if the people producing tech products were following the cardinal rule of drug dealing: never get high on your own supply.

This is unsettling. Why are the world's greatest public technocrats also its greatest private technophobes? Can you imagine the outcry if religious leaders refused to let their children practice religion? Many experts both within and beyond the world of tech have shared similar perspectives with me. Several video game designers told me they avoided the notoriously addictive game World of Warcraft; an exercise addiction psychologist called fitness watches dangerous—"the dumbest things in the world"—

and swore she'd never buy one; and the founder of an Internet addiction clinic told me she avoids gadgets newer than three years old. She has never used her phone's ringer, and deliberately "misplaces" her phone so she isn't tempted to check her email. (I spent two months trying to reach her by email, and succeeded only when she happened to pick up her office landline.) Her favorite computer game is Myst, released in 1993 when computers were still too clunky to handle video graphics. The only reason she was willing to play Myst, she told me, was because her computer froze every half hour and took forever to reboot.

Greg Hochmuth, one of Instagram's founding engineers, realized he was building an engine for addiction. "There's always another hashtag to click on," Hochmuth said. "Then it takes on its own life, like an organism, and people can become obsessive." Instagram, like so many other social media platforms, is bottomless. Facebook has an endless feed; Netflix automatically moves on to the next episode in a series; Tinder encourages users to keep swiping in search of a better option. Users benefit from these apps and websites, but also struggle to use them in moderation. According to Tristan Harris, a "design ethicist," the problem isn't that people lack willpower; it's that "there are a thousand people on the other side of the screen whose job it is to break down the self-regulation you have."

These tech experts have good reason to be concerned. Working at the far edge of possibility, they discovered two things. First, that our understanding of addiction is too narrow. We tend to think of addiction as something inherent in certain people—

those we label as *addicts*. Heroin *addicts* in vacant row houses. Chain-smoking nicotine *addicts*. Pill-popping prescription-drug *addicts*. The label implies that they're different from the rest of humanity. They may rise above their addictions one day, but for now they belong to their own category. In truth, addiction is produced largely by environment and circumstance. Steve Jobs knew this. He kept the iPad from his kids because, for all the advantages that made them unlikely substance addicts, he knew they were susceptible to the iPad's charms. These entrepreneurs recognize that the tools they promote—engineered to be irresistible—will ensnare users indiscriminately. There isn't a bright line between addicts and the rest of us. We're all one product or experience away from developing our own addictions.

Bilton's tech experts also discovered that the environment and circumstance of the digital age are far more conducive to addiction than anything humans have experienced in our history. In the 1960s, we swam through waters with only a few hooks: cigarettes, alcohol, and drugs that were expensive and generally inaccessible. In the 2010s, those same waters are littered with hooks. There's the Facebook hook. The Instagram hook. The porn hook. The email hook. The online shopping hook. And so on. The list is long—far longer than it's ever been in human history, and we're only just learning the power of these hooks.

Bilton's experts were vigilant because they knew they were designing irresistible technologies. Compared to the clunky tech of the 1990s and early 2000s, modern tech is efficient and addictive. Hundreds of millions of people share their lives in real time through Instagram posts, and just as quickly those lives are evaluated in the form of comments and likes. Songs that once took an hour to download now arrive in seconds, and the lag that dissuaded people from downloading in the first place has evaporated. Tech offers convenience, speed, and automation, but it also brings large costs. Human behavior is driven in part by a succession of reflexive cost-benefit calculations that determine whether an act will be performed once, twice, a hundred times, or not at all. When the benefits overwhelm the costs, it's hard not to perform the act over and over again, particularly when it strikes just the right neurological notes.

A like on Facebook and Instagram strikes one of those notes, as does the reward of completing a World of Warcraft mission, or seeing one of your tweets shared by hundreds of Twitter users. The people who create and refine tech, games, and interactive experiences are very good at what they do. They run thousands of tests with millions of users to learn which tweaks work and which ones don't—which background colors, fonts, and audio tones maximize engagement and minimize frustration. As an experience evolves, it becomes an irresistible, weaponized version of the experience it once was. In 2004, Facebook was fun; in 2016, it's addictive.

Addictive behaviors have existed for a long time, but in recent decades they've become more common, harder to resist, and more mainstream. These new addictions don't involve the ingestion of a substance. They don't directly introduce chemicals into your system, but they produce the same effects because they're compelling and well designed. Some, like gambling and exercise, are old; others, like binge-viewing and smartphone use, are relatively new. But they've all become progressively more difficult to resist.

Meanwhile, we've made the problem worse by focusing on

the benefits of goal-setting without considering its drawbacks. Goal-setting was a useful motivational tool in the past, because most of the time humans prefer to spend as little time and energy as possible. We're not intuitively hard-working, virtuous, and healthy. But the tide has turned. We're now so focused on getting more done in less time that we've forgotten to introduce an emergency brake.

I spoke to several clinical psychologists who described the magnitude of the problem. "Every single person I work with has at least one behavioral addiction," one psychologist told me. "I have patients who fit into every area: gambling, shopping, social media, email, and so on." She described several patients, all with high-powered professional careers, earning six figures, but deeply hobbled by their addictions. "One woman is very beautiful, very bright, and very accomplished. She has two master's degrees and she's a teacher. But she's addicted to online shopping, and she's managed to accumulate $80,000 in debt. She's managed to hide her addiction from almost everyone she knows." This compartmentalization was a common theme. "It's very easy to hide behavioral addictions—much more so than for substance abuse. This makes them dangerous, because they go unnoticed for years." A second patient, just as accomplished at work, managed to hide her Facebook addiction from her friends. "She went through a horrible breakup, and then stalked her ex-boyfriend online for years. With Facebook it's far more difficult to make a clean break when relationships end." A man she saw checked his email hundreds of times a day. "He's incapable of relaxing and enjoying himself on vacation. But you'd never know. He's deeply anxious, but he presents so well in the world; he has a successful career in the healthcare industry, and you'd never know how much he suffers."

"The impact of social media has been huge," a second psychologist told me. "Social media has completely shaped the brains of the younger people I work with. One thing I am often mindful of in a session is this: I could be five or ten minutes into a conversation with a young person about the argument they have had with their friend or girlfriend, when I remember to ask whether this happened by text, phone, on social media, or face-to-face. More often the answer is, 'text or social media.' Yet in their telling of the story, this isn't apparent to me. It sounds like what I would consider a 'real,' face-to-face conversation. I always stop in my tracks and reflect. This person doesn't differentiate various modes of communication the way I do . . . the result is a landscape filled with disconnection and addiction."

Irresistible traces the rise of addictive behaviors, examining where they begin, who designs them, the psychological tricks that make them so compelling, and how to minimize dangerous behavioral addiction as well as harnessing the same science for beneficial ends. If app designers can coax people to spend more time and money on a smartphone game, perhaps policy experts can also encourage people to save more for retirement or donate to more charities.

Technology is not inherently bad. When my brother and I moved with my parents to Australia in 1988, we left our grandparents in South Africa. We spoke to them once a week on expensive landline calls, and sent letters that arrived a week later.

When I moved to the United States in 2004, I emailed my parents and brother almost every day. We talked on the phone often, and waved to each other via webcam as often as we could. Technology shrank the distance between us. Writing for *Time* in 2016, John Patrick Pullen described how the emotional punch of virtual reality brought him to tears.

> . . . My playmate, Erin, shot me with a shrink ray. Suddenly, not only were all the toys enormous to me, but Erin's avatar was looming over me like a hulking giant. Her voice even changed as it poured through my headphones, entering my head with a deep, slow tone. And for a moment, I was a child again, with this giant person lovingly playing with me. It gave me such a profound perspective on what it must be like to be my son that I started to cry inside the headset. It was a pure and beautiful experience that will reshape my relationship with him moving forward. I was vulnerable to my giant playmate, yet felt completely safe.

Tech isn't morally good or bad until it's wielded by the corporations that fashion it for mass consumption. Apps and platforms can be designed to promote rich social connections; or, like cigarettes, they can be designed to addict. Today, unfortunately, many tech developments do promote addiction. Even Pullen, in rhapsodizing his virtual reality experience, said he was "hooked." Immersive tech like virtual reality inspires such rich emotions that it's ripe for abuse. It's still in its infancy, though, so it's too soon to know whether it will be used responsibly.

In many respects, substance addictions and behavioral addictions are very similar. They activate the same brain regions, and they're fueled by some of the same basic human needs: social engagement and social support, mental stimulation, and a sense of effectiveness. Strip people of these needs and they're more likely to develop addictions to both substances and behaviors.

Behavioral addiction consists of six ingredients: compelling goals that are just beyond reach; irresistible and unpredictable positive feedback; a sense of incremental progress and improvement; tasks that become slowly more difficult over time; unresolved tensions that demand resolution; and strong social connections. Despite their diversity, today's behavioral addictions embody at least one of those six ingredients. Instagram is addictive, for example, because some photos attract many likes, while others fall short. Users chase the next big hit of likes by posting one photo after another, and return to the site regularly to support their friends. Gamers play certain games for days on end because they're driven to complete missions, and because they've formed strong social ties that bind them to other gamers.

So what are the solutions? How do we coexist with addictive experiences that play such a central role in our lives? Millions of recovering alcoholics manage to avoid bars altogether, but recovering Internet addicts are forced to use email. You can't apply for a travel visa or a job, or begin working, without an email address. Fewer and fewer modern jobs allow you to avoid using computers and smartphones. Addictive tech is part of the mainstream in a way that addictive substances never will be. Abstinence isn't an option, but there are other alternatives. You can confine addictive experiences to one corner of your life, while courting good habits that promote healthy behaviors. Meanwhile, once you understand how behavioral addictions work, you can mitigate their harm, or even harness them for good. The same principles that drive children to play games might drive them to learn at school, and the goals that drive people to exercise addictively might also drive them to save money for retirement.

———

The age of behavioral addiction is still young, but early signs point to a crisis. Addictions are damaging because they crowd out other essential pursuits, from work and play to basic hygiene and social interaction. The good news is that our relationships with behavioral addiction aren't fixed. There's much we can do to restore the balance that existed before the age of smartphones, emails, wearable tech, social networking, and on-demand viewing. The key is to understand why behavioral addictions are so rampant, how they capitalize on human psychology, and how to defeat the addictions that hurt us, and harness the ones that help us.

PART 1

What Is Behavioral Addiction and
Where Did It Come From?

1.

The Rise of
Behavioral Addiction

A couple of years ago, Kevin Holesh, an app developer, decided that he wasn't spending enough time with his family. The culprit was technology, and his smartphone was the biggest offender. Holesh wanted to know how much time he was spending on his phone each day, so he designed an app called Moment. Moment tracked Holesh's daily screen time, tallying how long he used his phone each day. I spent months trying to reach Holesh because he lives by his word. On the Moment website, he writes that he may be slow to reply to email because he's trying to spend less time online. Eventually, after my third attempt, Holesh replied with a polite apology and agreed to talk. "The app stops tracking when you're just listening to music or making phone calls," Holesh told me. "It starts up again when you're looking at your screen—sending emails or browsing the web, for example." Holesh was spending an hour and fifteen minutes a day glued to his screen, which seemed like a lot. Some of his friends had similar concerns, but also had no idea how much time they lost to their phones. So Holesh shared the app. "I asked people to guess what their daily usage was and they were almost always 50 percent too low."

I downloaded Moment several months ago. I guessed I was using my phone for an hour a day at the most, and picking it up perhaps ten times a day. I wasn't proud of those numbers, but they sounded about right. After a month, Moment reported that I was using my phone for an average of three hours a day, and picking it up an average of forty times. I was stunned. I wasn't playing games or surfing the web for hours, but somehow I managed to spend twenty hours a week staring at my phone.

I asked Holesh whether my numbers were typical. "Absolutely," he said. "We have thousands of users, and their average usage time is just under three hours. They pick up their phones an average of thirty-nine times a day." Holesh reminded me that these were the people who were concerned enough about their screen time to download a tracking app in the first place. There are millions of smartphone users who are oblivious or just don't care enough to track their usage—and there's a reasonable chance they're spending even more than three hours on their phones each day.

Perhaps there was just a small clump of heavy users who spent all day, every day on their phones, dragging the average usage times higher. But Holesh shared the usage data of eight thousand Moment users to illustrate that wasn't the case at all:

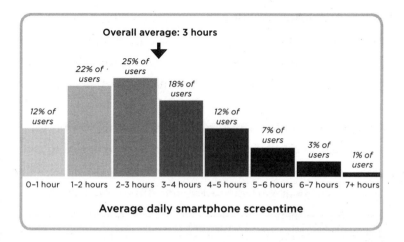

Overall average: 3 hours

12% of users

22% of users

25% of users

18% of users

12% of users

7% of users

3% of users

1% of users

0–1 hour   1–2 hours   2–3 hours   3–4 hours   4–5 hours   5–6 hours   6–7 hours   7+ hours

Average daily smartphone screentime

Most people spend between one and four hours on their phones each day—and many far longer. This isn't a minority issue. If, as guidelines suggest, we should spend less than an hour on our phones each day, 88 percent of Holesh's users were overusing. They were spending an average of a quarter of their waking lives on their phones—more time than any other daily activity, except sleeping. Each month almost one hundred hours was lost to checking email, texting, playing games, surfing the web, reading articles, checking bank balances, and so on. Over the average lifetime, that amounts to a staggering *eleven years*. On average they were also picking up their phones about three times an hour. This sort of overuse is so prevalent that researchers have coined the term "nomophobia" to describe the fear of being without mobile phone contact (an abbreviation of "no-mobile-phobia").

Smartphones rob us of time, but even their mere presence is damaging. In 2013, two psychologists invited pairs of strangers into a small room and asked them to engage in conversation. To smooth the process, the psychologists suggested a topic: why not discuss an interesting event that happened to you over the past month? Some of the pairs talked while a smartphone sat idle nearby, while for others the phone was replaced by a paper notebook. Every pair bonded to some extent, but those who grew acquainted in the presence of the smartphone struggled to connect. They described the relationships that formed as lower in quality, and their partners as less empathetic and trustworthy. Phones are disruptive by their mere existence, even when they aren't in active use. They're distracting because they remind us of the world beyond the immediate conversation, and the only solution, the researchers wrote, is to remove them completely.

Smartphones aren't the only culprits. Bennett Foddy has played thousands of video games, but refuses to play World of Warcraft. Foddy is a brilliant thinker with dozens of interests. He works as a game developer and professor at New York University's Game Center. Foddy was born and lived in Australia, where he was the bassist in an Australian band called Cut Copy— which released several best-selling singles and won a string of Australian music awards—until he moved, first to Princeton University and then to Oxford University, to study philosophy. Foddy has immense respect for WoW, as it's known, but won't play it himself. "I take it as part of my job to play all the culturally significant games. But I didn't play that one because I can't afford the loss of time. I know myself reasonably well, and I suspect it probably would have been difficult for me to shake."

WoW may be one of the most addictive behavioral experiences on the planet. It's a massively multiplayer online role-playing game, with millions of players from around the world who create avatars that roam across landscapes, fight monsters, complete quests, and interact with other players. Almost half of all players consider themselves "addicted." An article in *Popular Science* described WoW as "the obvious choice" when searching for the world's most addictive game. There are support groups with thousands of members, and more than a quarter of a million people have taken the free online World of Warcraft Addiction Test. In ten years, the game has grossed more than ten billion dollars, and attracted more than one hundred million subscribers. If they formed a nation, it would be the twelfth biggest on Earth. WoW players choose an avatar, which represents them as they complete quests in a virtual world called Azeroth. Many players band together to form guilds—teams of allied avatars—which is part of what makes the game so addictive. It's hard to sleep at night when you know three of your guild-mates in Copenhagen, Tokyo, and Mumbai are on an epic quest without you. As we chatted, I was struck by Foddy's passion for games. He believes without a doubt that they're a net force for good in the world—but still refuses to sample the charms of Azeroth for fear of losing months or years of his life.

Games like WoW attract millions of teens and young adults, and a considerable minority—up to 40 percent—develop addictions. Several years ago a computer programmer and a clinical psychologist joined forces to open a gaming and Internet addiction center in the woods near Seattle. The center, named reSTART, houses a dozen or so young men who are addicted to WoW, or one of a handful of other games. (reSTART tried admitting a small group of women, but many Internet addicts also develop sex addictions, so cohabitation became a major distraction.) Computers have never before had the memory to run games like WoW, which are much faster, more immersive, and less clunky than the games of the twentieth century. They allow you to interact with other people in real time, a huge part of what makes them so addictive.

Technology has also changed how we exercise. Fifteen years ago I bought an early model Garmin exercise watch, a mammoth rectangular device somewhere between a watch and a wrist weight. It was so heavy that I had to carry a water bottle in my other hand to balance its weight. It lost its GPS signal every couple of minutes, and battery life was so limited that it was useless on long runs. Today there are cheaper, smaller wearable devices that capture every step. That's miraculous, but also a recipe for obsession. Exercise addiction has become a psychiatric specialty because athletes are constantly reminded of their activity and, even more so, their inactivity. People who wear exercise watches become trapped in a cycle of escalation. Ten thousand steps may have been the gold standard last week, but this week it's eleven thousand. Next week, twelve thousand, and then fourteen thousand. That trend can't continue forever, but many people push through stress fractures and other major injuries to seek the same endorphin high that came from a much lighter exercise load only months earlier.

Intrusive tech has also made shopping, work, and porn harder to escape. It was once almost impossible to shop and work between the late evening and early morning, but now you can shop online and connect to your workplace any time of the day. Gone also are the days of stealing a copy of *Playboy* from the newsstand; all you need are Wi-Fi and a web browser. Life is more convenient than ever, but convenience has also weaponized temptation.

So how did we get here?

The first "behavioral addicts" were two-month-old babies. In early December 1968, forty-one psychologists who studied human vision met in New York City at the annual meeting of the Association for Research in Nervous and Mental Disease to discuss why our ability to see sometimes fails. It was a who's who of academic luminaries. Roger Sperry would win the Nobel Prize in medicine thirteen years later. Neuroscientist Wilder Penfield was once described as the "greatest living Canadian," and Stanford's William Dement was crowned "the father of sleep medicine."

In attendance was the psychologist Jerome Kagan, who a decade earlier had joined Harvard University to create the first program in human development. By his retirement half a century later, he was listed as the twenty-second most eminent psychologist of all time—ahead of giants like Carl Jung, Ivan Pavlov, and Noam Chomsky.

At the meeting, Kagan discussed visual attention in infants. How, he asked, do two-month-old babies know what to look at and what to ignore? Their growing brains are bombarded by a kaleidoscope of visual information, and yet somehow they learn to focus on some images and look past others. Kagan noticed that very young babies were drawn to moving, hard-edged objects. In fact, they couldn't look away when a researcher dangled a wooden block before them. According to Kagan, these infants were showing "a behavioral addiction to contour and movement."

By modern standards, though, it would be a stretch to call the infants behavioral addicts. Kagan was right that they couldn't look away, but the way we think of behavioral addiction today is quite different. It's more than an instinct that we can't override, because that would include blinking and breathing. (Try holding your breath till you pass out and your brain will eventually force you to breathe again. The fact that we can't help inhaling and exhaling means we're unlikely to die from forgetting to breathe.) Modern definitions recognize that addiction is ultimately a bad thing. A behavior is addictive only if the rewards it brings now are eventually outweighed by damaging consequences. Breathing and looking at wooden blocks aren't addictive because, even if they're very hard to resist, they aren't harmful. Addiction is a deep attachment to an experience that is harmful and difficult to do without. Behavioral addictions don't involve eating, drinking, injecting, or smoking substances. They arise when a person can't resist a behavior, which, despite addressing a deep psychological need in the short-term, produces significant harm in the long-term.

Obsession and *compulsion* are close relatives of behavioral addiction. Obsessions are thoughts that a person can't stop having, and compulsions are behaviors a person can't stop enacting. There's a key difference between addictions, and obsessions and compulsions. Addictions bring the promise of immediate reward, or positive reinforcement. In contrast, obsessions and compulsions are intensely unpleasant to *not* pursue. They promise relief—also known as negative reinforcement—but not the appealing rewards of a consummated addiction. (Since they're so closely related, I'll use all three terms in this book.)

Behavioral addiction also has a third relative in obsessive passion. In 2003, seven Canadian psychologists, led by the researcher Robert Vallerand, wrote a paper splitting the concept of passion in two. "Passion," they said, "is defined as a strong inclination toward an activity that people like, that they find important, and in which they invest time and energy." Harmonious passions are very healthy activities that people choose to do without strings attached—the model train set that an elderly man has been working on since his youth, or the series of abstract paintings that a middle-aged woman creates in her free time. "Individuals are not compelled to do the activity," the researchers said, "but rather they freely choose to do so. With this type of passion, the activity occupies a significant but not overwhelming space in the person's identity and is in harmony with other aspects of the person's life."

Obsessive passions, however, are unhealthy and sometimes dangerous. Driven by a need that goes beyond simple enjoyment, they're likely to produce behavioral addictions. As the researchers defined it, the individual "cannot help but to engage in the passionate activity. The passion must run its course as it controls the person. Because activity engagement is out of the person's control, it eventually takes disproportionate space in the person's identity and causes conflict with other activities in the person's life." This is the video game that a teenager plays all night instead of sleeping and doing his homework. Or the runner who once ran for fun, but now feels compelled to run at least six miles a day at a certain pace, even as debilitating stress injuries set in. Until she's on her back, unable to walk, she'll continue to run daily because her identity and well-being are intimately bound with her as yet unbroken streak. Harmonious passions "make life worth living," but an obsessive passion plagues the mind.

There are people, of course, who disagree with the idea that addictions can be purely behavioral. "Where are the substances?" they ask. "If you can be addicted to video games and smartphones, why can't you be addicted to smelling flowers or walking backward?" You *can* be addicted to those things, in theory. If they come to fulfill a deep need, you can't do without them, and you begin to pursue them while neglecting other aspects of your life, then you've developed a behavioral addiction to smelling flowers or walking backward. There probably aren't many people with those particular addictions, but they aren't inconceivable. Meanwhile, there are many, *many* people who show similar symptoms when you introduce them to a smartphone or a compelling video game or the concept of email.

There are also people who say that the term "addiction" can't possibly apply to a majority of the population. "Doesn't that devalue the term 'addiction'? Doesn't that make it meaningless and empty?" they ask. When, in 1918, a flu pandemic killed seventy-five million people, no one suggested that a flu diagnosis was meaningless. The issue demanded attention precisely because it affected so many people, and the same is true of behavioral addiction. Smartphones and email are hard to resist—because they're both part of the fabric of society and promote psychologically compelling experiences—and there will be other addictive experiences in the coming decades. We shouldn't use a watered-down term to describe them; we should acknowledge how serious they are, how much harm they're doing to our collective well-being, and how much attention they deserve. The evidence so far is concerning, and trends suggest we're wading deeper into dangerous waters.

Still, it's important to use the term "behavioral addiction" carefully. A label can encourage people to see a disorder everywhere. Shy kids were suddenly labeled "Asperger's sufferers" when the term became popular; people with volatile emotions were similarly labeled "bipolar." Allen Frances, a psychiatrist and expert on addiction, is concerned about the term "behavioral addiction." "If 35 percent of people suffer from a disorder, then it's just a part of human nature," he says. "Medicalizing behavioral addiction is a mistake. What we should be doing is what they do in Taiwan and Korea. There they see behavioral addiction as a social issue rather than a medical issue." I agree. Not everyone who uses a smartphone for more than ninety minutes a day should be in treatment. But what is it about smartphones that makes them so compelling? Should we introduce structural checks and balances on the growing role they play in our collective lives? A symptom affecting so many people is no less real or more acceptable simply because it becomes a new norm; we need to understand that symptom to decide whether and how to deal with it.

———

Just how common are behavioral addictions? The most debilitating addictions, which hospitalize people or render them incapable of living vaguely normal lives, are quite rare, affecting just a few percent of the population. But moderate behavioral addictions are far more common. These addictions make our lives less worthwhile, make us less effective at work and play, and diminish our interactions with other people. They inflict milder psychological traumas than severe addictions, but even milder traumas accumulate over time to degrade a person's well-being.

Figuring out how many people suffer from behavioral addictions is difficult, because most of these addictions go unreported. Dozens of studies have investigated the question, but the most comprehensive came from an English psychology professor named Mark Griffiths, who has been studying behavioral addiction for more than twenty years. He speaks as quickly and passionately as you'd expect from someone who has published more than five hundred papers midway through his career. A precocious student, Griffiths finished his doctorate at twenty-three—a couple of years before the Internet boom. "It was 1994," Griffiths says, "I was presenting a paper at an annual British Psychological Society meeting on technology and addiction, and there was a press conference after the talk. At that point people were talking about slot machine, video game, and TV addictions, and someone asked whether I'd heard about this new thing called the Internet, and whether it could lead to new types of addictions." At first Griffiths wasn't sure what to make of the Internet, but he was fascinated by the idea that it might be a route to addiction. He applied for government funding and began to study the topic.

Reporters often asked Griffiths how common behavioral addictions were, but he struggled to give them a definitive answer. The data just weren't available. So he joined forces with two researchers at the University of Southern California to find out. They published a long and thorough review paper in 2011, surveying dozens of studies, each carefully vetted before its inclusion. Studies were only included if they had at least five hundred respondents, both men and women, aged between sixteen and sixty-five years, and their measurement methods had to be reliable and supported with careful research. The result was an impressive eighty-three studies with a grand total of 1.5 million respondents from four continents. The studies focused on gambling, love, sex, shopping, Internet, exercise, and work addictions, as well as alcohol, nicotine, narcotic, and other substance addictions.

The bottom line: a staggering 41 percent of the population has suffered from at least one behavioral addiction over the past twelve months. These aren't trivial disorders; Griffiths and his colleagues were saying that almost half of the population had experienced the following symptoms:

> [The] loss of ability to choose freely whether to stop or continue the behavior (loss of control) and [the] experience of behavior-related adverse consequences. In other words, the

person becomes unable to reliably predict when the behavior will occur, how long it will go on, when it will stop, or what other behaviors may become associated with the addictive behavior. As a consequence, other activities are given up or, if continued, are no longer experienced as being as enjoyable as they once were. Further negative consequences of the addictive behavior may include interference with performance of life roles (e.g., job, social activities, or hobbies), impairment of social relationships, criminal activity and legal problems, involvement in dangerous situations, physical injury and impairment, financial loss, or emotional trauma.

Some of these addictions continue to grow with technological innovation and social change. One recent study suggested that up to 40 percent of the population suffers from some form of Internet-based addiction, whether to email, gaming, or porn. Another found that 48 percent of its sample of U.S. university students were "Internet addicts," and another 40 percent were borderline or potential addicts. When asked to discuss their interactions with the Internet, most of the students gravitated toward negative consequences, explaining that their work, relationship, and family lives were poorer because they spent too much time online.

At this point, you may be wondering whether you or someone you love is technically "addicted to the Internet." This is a sample of five questions from the twenty-item Internet Addiction Test, a widely used measure of Internet addiction. Take a moment to answer each question using the scale below, from 0 to 5:

Internet Addiction Test

Select the response that best represents the frequency of each behavior listed using the scale below:

0 = Not applicable
1 = Rarely
2 = Occasionally
3 = Frequently
4 = Often
5 = Always

How often do you find that you stay online longer than you intended? ____

How often do others in your life complain to you about the amount of time you spend online? ___

How often do you check your email before something else that you need to do? ___

How often do you lose sleep because of late night log-ins? ____

How often do you find yourself saying "just a few minutes" when online? ___

If you scored 7 or below, you show no signs of Internet addiction. A score of 8–12 suggests mild Internet addiction—you may spend too long on the web sometimes, but you're generally in control of your usage. A score of 13–20 indicates moderate Internet addiction, which implies that your relationship with the Internet is causing you "occasional or frequent problems." A score between 21 and 25 suggests severe Internet addiction, and implies that the Internet is causing "significant problems in your life." (I'll return to the question of how to deal with a high score in the third section of this book.)

Beyond Internet addiction, 46 percent of people say they couldn't bear to live without their smartphones (some would rather suffer physical injury than an injury to their phones), and 80 percent of teens check their phones at least once an hour. In 2008, adults spent an average of eighteen minutes on their phones per day; in 2015, they were spending two hours and forty-eight minutes per day. This shift to mobile devices is dangerous, because a device that travels with you is always a better vehicle for addiction. In one study, 60 percent of respondents reported binge-watching dozens of television episodes in a row despite planning to stop much sooner. Up to 59 percent of people say they're dependent on social media sites and that their reliance on these sites ultimately makes them unhappy. Of that group, half say they need to check those sites at least once an hour. After an hour, they are anxious, agitated, and incapable of concentrating. Meanwhile, in 2015, there were 280 million smartphone addicts. If they banded together to form the "United States of Nomophobia," it would be the fourth most populous country in the world, after China, India, and the United States.

In 2000, Microsoft Canada reported that the average human had an attention span of twelve seconds; by 2013 that number had fallen to eight seconds. (According to Microsoft, a goldfish, by comparison, has an average attention span of nine seconds.) "Human attention is dwindling," the report declared. Seventy-seven percent of eighteen- to twenty-four-year-olds claimed that they reached for their phones before doing anything else when nothing is happening. Eighty-seven percent said they often zoned out, watching TV episodes back-to-back. More worrying, still, Microsoft asked two thousand young adults to focus their attention on a string of numbers and letters that appeared on a computer screen. Those who spent less time on social media were far better at the task.

———

Addiction originally meant a different kind of strong connection: in ancient Rome, being addicted meant you had just been sentenced to slavery. If you owed someone money and couldn't repay the debt, a judge would sentence you to addiction. You'd be forced to work as a slave until you'd repaid your debt. This was the first use of the word *addiction*, but it evolved to describe any bond that was difficult to break. If you liked to drink wine, you were a wine addict; if you liked to read books, you were a book addict. There was nothing fundamentally wrong with being an addict; many addicts were just people who *really* liked eating or drinking or playing cards or reading. To be an addict was to be passionate about something, and the word *addiction* became diluted over the centuries.

In the 1800s, the medical profession breathed new life into the word. In particular, doctors paid special attention in the late 1800s when chemists learned to synthesize cocaine, because it became more and more difficult to wean users off the drug. At first cocaine seemed like a miracle, allowing the elderly to walk for miles and the exhausted to think clearly again. In the end, though, most users became addicted, and many failed to survive.

I'll return to behavioral addiction shortly, but to understand its rise I'll need to focus on substance addiction first. The word "addiction" has only implied substance abuse for two centuries, but hominids have been addicted to substances for thousands of years. DNA evidence suggests that Neanderthals carried a gene known as DRD4-7R as long as forty thousand years ago. DRD4-7R is responsible for a constellation of behaviors that set Neanderthals apart from earlier hominids, including risk-taking, novelty-seeking, and sensation-seeking. Where pre-Neanderthal hominids were timid and risk-averse, Neanderthals were constantly exploring and rarely satisfied. A variant of DRD4-7R known as DRD4-4R is still present in about 10 percent of the population, who are far more likely than others to be daredevils and serial addicts.

It's impossible to pinpoint the first human addict, but records suggest he or she lived more than thirteen thousand years ago. The world was a very different place then. Neanderthals were long extinct, but the Earth was still covered in glaciers, the woolly mammoth would exist for another two thousand years, and humans were just beginning to domesticate sheep, pigs, goats, and cows. Farming and agriculture would only begin several millennia later, but on the Southeast Asian island of Timor, someone stumbled onto the betel nut.

The betel nut is the ancient, unrefined cousin of the modern cigarette. Betel nuts contain an odorless oily liquid known as arecoline, which acts much like nicotine. When you chew a betel nut, your blood vessels dilate, you breathe more easily, your blood pumps faster, and your mood lightens. People often claim to think more clearly after chewing a betel nut, and it's still a popular drug of choice in parts of South and Southeast Asia.

Betel nuts, however, have a nasty side effect. If you chew them often enough, your teeth will become black and rotten and they may fall out. Despite the obvious cosmetic costs of chewing the nuts, plenty of users continue chewing even as they lose their teeth. When Chinese emperor Zhou Zhengwang visited Vietnam two thousand years ago, he asked his hosts why their teeth were black. They explained that "betel-chewing is for keeping good sanitary conditions in the mouth; therefore, teeth turn black." This is shaky logic, at best. When parts of you turn pitch black, you need an open mind to conclude that the transformation is healthy.

South Asians weren't the only ancient addicts. Other civilizations delved into whatever grew locally. For thousands of years, residents of the Arabian Peninsula and the Horn of Africa have been chewing the khat leaf, a stimulant that acts like the drug speed, or methamphetamine. Khat users become talkative, euphoric, and hyperactive, and their heart rates rise as though they've had several cups of strong coffee. Around the same time, Aboriginal Australians stumbled upon the pituri plant, while their contemporaries in North America discovered the tobacco plant. Both plants can be smoked or chewed, and both contain heavy doses of nicotine. Meanwhile, seven thousand years ago, South Americans in the Andes began chewing the leaves of the coca plant at large communal gatherings. A hemisphere away, the Samarians were learning to prepare opium, which pleased them so much that they etched instructions on small clay tablets.

Substance addiction, as we know it, is relatively new, because it relies on sophisticated chemistry and expensive equipment. In television's *Breaking Bad*, chemistry-teacher-turned-meth-cook Walter White is obsessed with the purity of his product. He produces "Blue Sky," which is 99.1 percent pure, and earns immense global respect (and millions of dollars in drug money). But, in reality, meth addicts will buy anything they can find, so meth dealers cut the raw product with fillers that dilute its purity. Regardless of the emphasis on purity, the process of manufacturing the drug is intricate and technical. The same is true of many other drugs, which are chemically quite different from the raw plants that contain their primary ingredients.

Before drugs were big business, doctors and chemists discovered their effects by trial and error, or by accident. In 1875 the British Medical Association elected seventy-eight-year-old Sir Robert Christison as its forty-fourth president. Christison was tall, severe, and eccentric. He had begun practicing medicine fifty years earlier, just as homicidal Englishmen were learning to poison each other with arsenic, strychnine, and cyanide. Christison wondered how these and other toxins affected the human body. Volunteers were hard to come by, so he spent decades swallowing and regurgitating dangerous poisons himself, documenting their effects in real time just before he lost consciousness.

One of those toxins was a small green leaf, which numbed Christison's mouth, gave him a burst of long-lasting energy, and left him feeling decades younger than his eighty years. Christison was so invigorated that he decided to set out for a long walk. Nine hours and fifteen miles later he returned home and wrote that he was neither hungry nor thirsty. The next morning, he awoke feeling fit and ready to tackle the new day. Christison had been chewing on the coca leaf, the plant responsible for its famous stimulant cousin, cocaine.

In Vienna, one thousand miles to the southeast, a young neurologist was also experimenting with cocaine. Many people remember Sigmund Freud for his theories of human personality, sexuality, and dreaming, but he was also famous in his day for promoting cocaine. Chemists had first synthesized the drug three decades earlier, and Freud read of Christison's miraculous fifteen-mile stroll with interest. Freud found that cocaine not only gave him energy, but also calmed his recurring bouts of depression and indigestion. In one of more than nine hundred letters to his fiancée, Martha Bernays, Freud wrote:

> If it goes well I will write an essay on [cocaine] and I expect it will win its place in therapeutics by the side of morphium and superior to it . . . I take very small doses of it regularly against depression and against indigestion, and with the most brilliant success.

Freud's life was filled with highs and lows, but the decade that followed this letter to Martha was particularly turbulent. It began with a high point: the publication of his essay titled "Über Coca" in 1884. In Freud's words, "Über Coca" was "a song of praise to this magical substance." Freud played every part in the "Über Coca" drama; he was experimenter, research subject, and animated writer.

> A few minutes after taking cocaine, one experiences a sudden exhilaration and feeling of lightness. One feels a certain furriness on the lips and palate, followed by a feeling of warmth in the same areas . . . The psychic effect of

[cocaine] . . . consists of exhilaration and lasting euphoria, which does not differ in any way from the normal euphoria of a healthy person.

"Über Coca" also hints at cocaine's darker side, though Freud seemed more fascinated than concerned:

During this first trial I experienced a short period of toxic effects . . . Breathing became slower and deeper and I felt tired and sleepy; I yawned frequently and felt somewhat dull . . . If one works intensively while under the influence of coca, after from three to five hours there is a decline in the feeling of well-being, and a further dose of coca is necessary in order to ward off fatigue.

Many psychologists have criticized Freud because his most famous theories are impossible to test (are men who dream of caves really preoccupied with the womb?), but he championed careful experimentation with cocaine. As his letters show, Freud discovered that cocaine, like any addictive stimulus, wore off and its effects weakened over time. The only way to recreate the original high was with repeated, escalating doses. He took at least a dozen large doses, and ultimately became addicted. He struggled to think and work without the drug, and became convinced his best ideas flowered under its influence. In 1895, his nose became infected and he endured operations to repair his collapsed nostrils. In one letter to his friend and ear, nose, and throat specialist, Wilhelm Fliess, Freud described in graphic detail the effects of cocaine. Ironically, the only thing that soothed his nose was another dose of cocaine. When the pain was particularly bad, he painted his nostrils with a solution of water and cocaine. A year later, dejected, he concluded that cocaine was more harmful than helpful. In 1896, twelve years after first encountering cocaine, Freud was forced to stop using the drug completely.

How could Freud see cocaine's upside but not its staggering downside? Early in his infatuation with the drug, he decided it was the answer to morphine addiction. He described the case of a patient who quit morphine cold turkey and went into "sudden withdrawal," wracked by chills and bouts of depression. But when the man began ingesting cocaine, he recovered completely, functioning normally with the help of a heavy daily dose of cocaine. Freud's biggest mistake was to believe that this effect was permanent:

> After ten days he was able to dispense with the coca treatment altogether. The treatment of morphine addiction with coca does not, therefore, result merely in the exchange of one kind of addiction for another . . . the use of coca is only temporary.

Freud was seduced by cocaine in part because he lived during a time when addiction was presumed to affect people who were weak of mind and body. Genius and addiction were incompatible, and he (like Robert Christison) discovered cocaine at the height of his intellectual powers. Freud so deeply misunderstood the drug that he believed it could replace and eliminate morphine addiction. He wasn't the only person to hold this belief. Two decades before Freud wrote "Über Coca," a Confederate Army colonel became addicted to morphine after he was injured during the final battle of the American Civil War. He, too, believed he could overcome his morphine addiction with a cocaine-laced tincture. He was wrong, but his medicine ultimately became one of the most widely consumed substances on Earth.

The Civil War ended with a brief but bloody battle on the evening of Easter, April 16, 1865. The Union and Confederate armies converged on the Chattahoochee River, near Columbus, Georgia, and fought on horseback near two bridges that spanned the river. One unfortunate Confederate soldier, John Pemberton, encountered a wall of Union cavalrymen when he tried to block a bridge that led into the heart of Columbus. Pemberton brandished a saber, but before he could use it he was shot. As he reared back in agony, a Union soldier inflicted a deep slash across Pemberton's chest and stomach. He slumped down, near death, but was dragged to safety by a friend.

Pemberton survived, but his saber wound burned for months. Like thousands of other injured soldiers, he treated his pain with morphine. At first, army doctors administered small doses spread many hours apart, but Pemberton began to tolerate the drug. He demanded bigger doses more and more often, and eventually developed a full-blown addiction. The doctors did their best to wean him off the drug, but they were undermined at every step—Pemberton had been a chemist before the war, so his old suppliers stepped in when the army's contribution dwindled. His friends became concerned, and Pemberton was ultimately forced to acknowledge that morphine was doing his body more harm than good.

Like any good scientist—and like Freud after him—Pemberton experimented. His goal was a non-addictive replacement for morphine to relieve chronic pain. By the 1880s, after several false starts, Pemberton found a winner in Pemberton's French Wine Coca: a combination of wine, coca leaves, kola nuts, and an aromatic shrub called damiana. There was no Food and Drug Administration in the 1880s, so Pemberton was free to wax lyrical (and ungrammatical) about the tonic's medical properties, even if he wasn't quite sure how it worked. He paid for one newspaper ad in 1885, which read:

French Wine Coca is indorsed by over 20,000 of the most learned and scientific medical men in the world . . .

. . . Americans are the most nervous people in the world . . . All who are suffering from any nervous complaints we commend to use the wonderful and delightful remedy, French Wine Coca, infallible in curing all who are afflicted with any nerve trouble, dyspepsia, mental and physical exhaustion, all chronic wasting diseases, gastric irritability, constipation, sick headache, neuralgia, etc. is quickly cured by the Coca Wine . . .

. . . Coca is a most wonderful invigorator of the sexual organs and will cure seminal weakness, impotency, etc., when all other remedies fail . . .

To the unfortunate who are addicted to the morphine or opiate habit, or the excessive use of alcohol stimulants, the

French Wine Coca has proven a great blessing, and thousands proclaim it the most remarkable invigorator that every sustained a wasting and sinking system.

Like Sigmund Freud, Pemberton believed that a combination of caffeine and coca leaves would conquer his morphine addiction without introducing a new one in its place. When the local government introduced prohibition laws in 1886, Pemberton removed the wine from his medicine, rechristening it Coca-Cola.

The story splits in two here. For the product, Coca-Cola, the sky was the limit. Coca-Cola went from strength to strength, sold first to business tycoon Asa Candler, and then to marketing geniuses Ernest Woodruff and W. C. Bradley. Woodruff and Bradley devised the brilliant idea of selling Coke in six-packs, to be carried more easily between the store and home, and both became immeasurably rich. For the man, John Pemberton, the opposite was true. Coca-Cola turned out not to be a viable replacement for morphine, and his addiction deepened. Instead of replacing morphine, cocaine compounded the problem, Pemberton's health continued to decline, and in 1888, he died penniless.

It's easy to look back at how little Freud and Pemberton understood of cocaine with a sense of superiority. We teach our children that cocaine is dangerous, and it's hard to believe that experts considered the drug a panacea only a century ago. But perhaps our sense of superiority is misplaced. Just as cocaine charmed Freud and Pemberton, today we're enamored of technology.

We're willing to overlook its costs for its many gleaming benefits: for on-demand entertainment portals, car services, and cleaning companies; Facebook and Twitter; Instagram and Snapchat; Reddit and Imgur; Buzzfeed and Mashable; Gawker and Gizmodo; online gambling sites, Internet video platforms, and streaming music hubs; hundred-hour work weeks, power naps, and four-minute gym workouts; and the rise of a new breed of obsessions, compulsions, and addictions that barely existed during the twentieth century.

And then there's the social world of the modern teen.

In 2013, a psychologist named Catherine Steiner-Adair explained that many American children first encounter the digital world when they notice that their parents are "missing in action." "My mom is almost always on the iPad at dinner," a seven-year-old named Colin told Steiner-Adair. "She's always 'just checking.'" Penny, also seven, said, "I always keep on asking her let's play let's play and she's always texting on her phone." At thirteen, Angela wished her parents understood "that technology isn't the whole world . . . it's annoying because it's like *you also have a family! How about we just spend some time together*, and they're like, 'Wait, I just want to check something on my phone. I need to call work and see what's going on.' Parents with younger kids do even more damage when they constantly check their phones and tablets. Using head-mounted cameras, researchers have shown that infants instinctively follow their parents' eyes. Distracted parents cultivate distracted children, because parents who can't focus teach their children the same attentional patterns. According to the paper's lead researcher, "The ability of children to sustain attention is known as a strong indicator for later success in areas such as language acquisition, problem-solving, and other key cognitive development milestones. Caregivers who appear distracted or whose eyes wander a lot while their children play appear to negatively impact infants' burgeoning attention spans during a key stage of development."

Kids aren't born craving tech, but they come to see it as indispensable. By the time they enter middle school, their social lives migrate from the real world to the digital world. All day, every day, they share hundreds of millions of photos on Instagram and billions of text messages. They don't have the option of taking a break, because this is where they come for validation and friendship.

Online interactions aren't just different from real-world interactions; they're measurably worse. Humans learn empathy and understanding by watching how their actions affect other people. Empathy can't flourish without immediate feedback, and it's a very slow-developing skill. One analysis of seventy-two studies found that empathy has declined among college students between 1979 and 2009. They're less likely to take the perspective of other people, and show less concern for others. The problem is bad among boys, but it's worse among girls. According to one study, one in three teenage girls say that people their age are mostly unkind to one another on social network sites. That's true for one in eleven boys aged twelve to thirteen, and one in six boys aged fourteen to seventeen.

Many teens refuse to communicate on the phone or face-to-face, and they conduct their fights by text. "It's too awkward in person," one girl told Steiner-Adair. "I was just in a fight with someone and I was texting them, and I asked, 'Can I call you, or can we video-chat?' and they were like, 'No.'" Another girl said, "You can think it through more and plan out what you want to say, and you don't have to deal with their face or see their reaction." That's obviously a terrible way to learn to communicate, because it discourages directness. As Steiner-Adair said, "Texting is the worst possible training ground for anyone aspiring to a mature, loving, sensitive relationship." Meanwhile, teens are locked into this medium. They either latch onto the online world, or they choose not to "spend time" with their friends.

Like Steiner-Adair, journalist Nancy Jo Sales interviewed girls aged between thirteen and nineteen to understand how they interacted with social media. For two and a half years she traveled around the United States, visiting ten states and speaking to hundreds of girls. She, too, concluded that they were enmeshed in the online world, where they learned and encountered cruelty, oversexualization, and social turmoil. Sometimes social media was just another way to communicate—but for many of the girls, it was a direct route to heartache. As addictive contexts go, this was a perfect storm: almost every teenage girl was using one or more social media platforms, so they were forced to choose between social isolation and compulsive overuse. No wonder so many of them spent hours texting and uploading Instagram posts every day after school; by all accounts, that was the rational thing to do. Echoing Sales's account, Jessica Contrera wrote a piece called "13, Right Now" for the *Washington Post*. Contrera chronicled several days in the life of a thirteen-year-old named Katherine Pommerening, a regular eighth grader who lumbered beneath the weight of so many "likes and lols." The saddest quote from Pommerening herself comes near the end of the article: "I don't feel like a child anymore," Katherine says. "I'm not doing anything childish. At the end of sixth grade"— when all her friends got phones and downloaded Snapchat, Instagram, and Twitter—"I just stopped doing everything I normally did. Playing games at recess, playing with toys, all of it, done."

Boys spend less time engaged in damaging online interactions, but many of them are hooked on games instead. The problem is so visible that some game developers are pulling their games from the market. They've begun to feel remorseful—not because their games feature sex or violence, but because they're devilishly addictive. With just the right combination of anticipation and feedback, we're encouraged to play for hours, days, weeks, months, and years at a time. In May 2013, a reclusive Vietnamese video game developer named Dong Nguyen released a game called Flappy Bird. The simple smartphone game asked players to guide a flying bird through obstacles by repeatedly tapping their phone screens. For a while, most gamers ignored Flappy Bird, and reviewers condemned the game because it was too difficult and seemed too similar to Nintendo's Super Mario Bros. For eight months Flappy Bird languished at the bottom of the app download charts.

But Nguyen's fortune changed in January 2014. Flappy Bird

attracted thousands of downloads overnight, and by the end of the month, the game was the most downloaded free app at Apple's online store. At the game's peak, Nguyen's design studio was earning $50,000 a day from ad revenue alone.

For a small-time game designer, this was the Holy Grail. Nguyen should have been ecstatic, but he was torn. Dozens of reviewers and fans complained that they were hopelessly addicted to Flappy Bird. According to Jasoom 79 on the Apple store website, "It ruined my life . . . its side effects are worse than cocaine/meth." Walter19230 titled his review "The Apocalypse," and began "My life is over." Mxndlsnsk warned prospective gamers not to download the game: "Flappy Bird will be the death of me. Let me start by saying DO NOT download Flappy Bird . . . People warned me, but I didn't care . . . I don't sleep, I don't eat. I'm losing friends."

Even if the reviews were exaggerated, the game seemed to be doing more harm than good. Hundreds of gamers made Nguyen sound like a drug dealer when they compared his product to meth and cocaine. What began as an idealistic labor of love appeared to be corrupting lives, and Nguyen's conscience overshadowed his success. On February 8, 2014, he tweeted:

I am sorry 'Flappy Bird' users, 22 hours from now, I will take 'Flappy Bird' down. I cannot take this anymore.

Some Twitter users believed Nguyen was responding to intellectual property claims, but he quickly dismissed that assumption:

It is not anything related to legal issues. I just cannot keep it anymore.

The game disappeared on cue and Nguyen evaded the limelight. Hundreds of Flappy Bird imitations popped up online, but Nguyen was already focused on his next project—a more complex game that was specifically designed *not* to be addictive.

Flappy Bird was addictive in part because everything about the game moved fast: the finger taps, the time between games, the onslaught of new obstacles. The world beyond Flappy Bird also moves faster than it used to. Sluggishness is the enemy of addiction, because people respond more sharply to rapid links between action and outcome. Very little about our world today— from technology to transport to commerce—happens slowly, and so our brains respond more feverishly.

Addiction is today better understood than in the nineteenth century, but it has also morphed and changed over time. Chemists have concocted dangerously addictive substances, and the entrepreneurs who design experiences have concocted similarly addictive behaviors. This evolution has only accelerated over the past two or three decades, and shows no signs of slowing. Just recently a doctor identified the first Google Glass addict— an enlisted naval officer who developed withdrawal symptoms when he tried to wean himself off the gadget. He'd been using it for eighteen hours a day, and he began to experience his dreams as though he were looking through the device. He'd managed to overcome alcohol addiction, he told doctors, but this was much worse. At night, when he relaxed, his right index finger would repeatedly float up to the side of his face. It was searching for the Glass power button, which was no longer there.

2.

The Addict in All of Us

Most war films ignore the boredom that sets in between bouts of action. In Vietnam, thousands of American G.I.s spent weeks, months, or even years just waiting. Some waited for instructions from their superiors, others for the action to arrive. Hugh Penn, a Vietnam vet, recalled that G.I.s passed the time by playing touch football and drinking beer at $1.85 per case. But boredom is the natural enemy of good behavior, and not everyone took to wholesome, all-American pastimes.

Vietnam lies just outside a region of Southeast Asia known as the Golden Triangle. This region encompasses Myanmar, Laos, and Thailand, and was responsible for most of the world's heroin supply during the Vietnam War. Heroin comes in different grades, and most Golden Triangle labs at the time were producing a chunky, low-grade product known as no. 3 heroin. In 1971, that all changed. The labs invited a series of master chemists from Hong Kong who had perfected a dangerous process known as ether precipitation. They started turning out no. 4 heroin, which was up to 99 percent pure. As the price of heroin rose from $1,240 to $1,780 per kilo, it began to find its way to South Vietnam, where bored G.I.s were just waiting to be entertained.

Suddenly, no. 4 heroin was everywhere. Teenage girls sold vials from roadside stands along the highway between Saigon and the Long Binh U.S. army base. In Saigon, street merchants crammed sample vials into the pockets of passing G.I.s, hoping they would return later for a second dose. The maids who cleaned the army barracks sold vials as they worked. In interviews, 85 percent of the returning G.I.s said they had been offered heroin. One soldier was offered heroin as he disembarked from the plane that brought him to Vietnam. The salesman, a heroin-addled soldier returning home from the war, asked only for a sample of urine so he could convince the U.S. authorities that he was clean.

Few of these soldiers had been within a mile of heroin before joining the army. They arrived healthy and determined to fight, but now they were developing addictions to some of the strongest stuff on the planet. By the war's end, 35 percent of the enlisted men said they had tried heroin, and 19 percent said they were addicted. The heroin was so pure that 54 percent of all users became addicted—many more than the 5–10 percent of amphetamine and barbiturate users who developed addictions in Vietnam.

Word of the epidemic filtered back to Washington, where government bureaucrats were forced to act. In early 1971, President Richard Nixon sent two U.S. congressmen to Vietnam to

gauge the epidemic's severity. The congressmen, Republican
Robert Steele and Democrat Morgan Murphy, rarely saw eye to
eye, but they agreed it was a catastrophe. They discovered that
ninety enlisted men had died from heroin overdoses in 1970, and
expected the numbers to rise by the close of 1971. Both men were
approached by heroin vendors during their short stay in Saigon,
and they were convinced the drug would find its way back to the
United States. "The Vietnam War is truly coming home to haunt
us," Steele and Morgan said in a report. "The first wave of heroin
is already on its way to our children in high school." The *New
York Times* printed an enlarged photo of Steele with a vial of heroin in his hand to show how easy it was for G.I.s to access the
drug. A *Times* editorial piece argued for the withdrawal of all
U.S. troops from Vietnam "to save the country from a debilitating drug epidemic."

At a press conference on June 17, 1971, President Nixon an-
nounced a war on drugs. He looked into the cameras with grave
determination, and said, "Public enemy number one in the
United States is drug abuse."

―――――

Nixon and his aides were worried, not just because the soldiers
were addicted to heroin in Vietnam, but about what would
happen when they returned home. How do you deal with a sudden influx of 100,000 heroin addicts? The problem was all the
worse because heroin was the most insidious drug on the market.

When British researchers assessed the harm of various drugs,
heroin was the worst by a big margin. On three scales measuring
the likelihood that a drug would inflict physical harm, induce

addiction, and cause social harm, heroin scored the highest rating—three out of three. It was by far the most dangerous and addictive drug in the world.

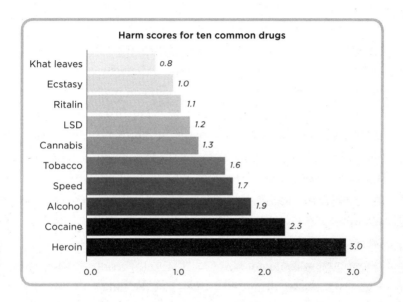

Harm scores for ten common drugs

Drug	Score
Khat leaves	0.8
Ecstasy	1.0
Ritalin	1.1
LSD	1.2
Cannabis	1.3
Tobacco	1.6
Speed	1.7
Alcohol	1.9
Cocaine	2.3
Heroin	3.0

It was hard enough to wean heroin addicts off the drug, but 95 percent relapsed at least once even after they'd detoxed. Few ever completely kicked the habit. Nixon was right to worry. He put together a team of experts who spent every waking hour planning for the onslaught of 100,000 new rehab patients. Nixon's team decided that the addicted G.I.s should stay in Vietnam until they were clean.

The government settled on a two-pronged attack, bolstering resources in Vietnam and at home. In Vietnam, Major General John Cushman was charged with cracking down on heroin use, which was so widespread that Cushman could see the problem by walking through camp. Doctors confirmed that hundreds if not thousands of men were addicted to the drug. Shocked by the extent of the problem, Cushman pursued a crackdown. At 5:30 one morning, he surprised his troops by confining them to base for twenty-four hours. Everyone was searched, and emergency medical clinics were set up to treat users as they detoxed. Heroin was so hard to come by that desperate users were forced to pay forty dollars per vial—up from three dollars per vial just a day earlier. At first Cushman seemed to have the upper hand, as three hundred men turned themselves in for treatment. But days later, as soon as he relaxed the travel ban, usage rates spiked again. Within a week heroin was selling for four dollars per vial, and more than half of the men who tried to detox were back on the drug.

At home, the government appointed a researcher named Lee Robins to monitor the progress of returning soldiers. Robins was a professor of psychiatry and sociology at Washington University, in St. Louis, where she studied the root causes of psychiatric epidemics. Robins was known for her uncanny ability to ask the right interview questions at just the right time. People trusted her, and she seemed to uncover sensitive information that interview subjects usually preferred not to share. The government decided that Robins was the perfect person to interview and track the re-covery of thousands of addicted G.I.s as they returned home.

For Robins this was an extraordinary opportunity. "[Study-ing] heroin use in a highly exposed normal population was unique," she reflected in 2010, "because there is nowhere else in the world where heroin is commonly used."

In the United States itself, heroin use is so rare that [a national survey] of 2,400 adults obtained only about 12 people who had used heroin in the last year. Because heroin users are scarce both worldwide and in the United States, most of our information about heroin comes from treated criminal samples.

But when Robins began following the returning vets, she was confused. What she found made no sense at all.

Normally just 5 percent of all heroin addicts stay clean, but Robins found that only 5 percent of the recovering G.I.s *relapsed*. Somehow, 95 percent managed to stay clean. The public, waiting for a calamity after Nixon's high-profile press conference, was naturally convinced that Robins was hiding the truth. Robins spent years defending the study. She wrote papers with headings like "Why the study was a technical success," and "The study's assets." Her detractors asked her, over and over, how she could be sure her results were accurate, and, if they were accurate, why so few of the G.I.s used the drug after they had returned home. It's easy to understand their skepticism. She had been appointed by a beleaguered president who declared a war on drugs, and Robins's report suggested he was gaining the upper hand. Even if she had been above politics, her results were simply too good to be true. In the world of public health, victories take the form of incremental reductions—a 3 percent drop here or a 5 percent drop there. A 90 percent drop in relapse rates was outlandish. But Robins had done everything right. Her experiment was sound and the results were real. The problem was explaining *why* only 5 percent of her G.I. subjects had relapsed.

The answer, it turned out, had been uncovered more than a decade earlier in a U.S. neuroscience lab some eight thousand miles away.

G reat scientists make their discoveries using two distinct approaches: tinkering and revolutionizing. Tinkering slowly wears down a problem, like water erodes rock, whereas in revolutions, a great thinker sees what no one else can. If the engineer Peter Milner was a tinkerer, the psychologist James Olds was a revolutionary. Together they made a superb team. In the early 1950s, in a small basement lab filled with caged rats and electrical equipment at Montreal's McGill University, Olds and Milner ran one of the most famous addiction experiments of all time. What made the experiment so remarkable was that it wasn't actually designed to reshape our understanding of addiction.

In fact, it might have gone unnoticed if Olds had done his job properly.

Olds and Milner met at Montreal's McGill University in the early 1950s. In many ways they were opposites. Milner's biggest strength was his technical know-how. He knew all there was to know about rat brains and electrical currents. Olds, on the other hand, lacked experience but overflowed with big ideas. Young researchers floated in and out of Olds' lab, drawn to his flair and talent for spotting the next big thing. Bob Wurtz, Olds' first graduate student in the late 1950s, knew Olds and Milner well. According to Wurtz, "Olds didn't know the front of the rat from the back of the rat, and Milner's first job was to educate Olds on rat physiology." But what Olds lacked in technical prowess, he more than made up for with brio and vision. "Jim was a very aggressive scientist," says Wurtz. "He believed in serendipity—if you see something interesting, you drop everything else. Whenever he and Milner stumbled on something newsworthy, Jim would deal with the media while Milner continued working in the lab."

Gary Aston-Jones, who also studied with Olds, remembered him the same way. "Olds was focused on big questions. He was always more conceptually driven than technically driven. When we were trying to understand how a fruit fly could learn about the world, Olds dropped to his hands and knees, crawled around on the floor, and pretended to be a fly." Milner would never have approached the problem that way. Aryeh Routtenberg, a third student who worked with Olds, explained that "Milner was sort of like the other face of Olds. He was quiet, humble, and self-effacing, while Olds would proclaim 'we've made a big discovery!'"

For decades, experts had assumed that drugs addicts—laudanum lushes, poppy tea drinkers, and opiate addicts—were predisposed to the condition, somehow wired incorrectly. Olds and Milner were some of the first researchers to turn that idea on its head—to suggest that, perhaps, under the right circumstances, we could all become addicts.

Their biggest discovery began modestly. Olds and Milner were trying to show that rats would run to the far end of their cages whenever an electric current zapped their tiny brains. The researchers implanted a small probe, which delivered a burst of electric current to each rat's brain when the rat pressed a metal bar. To their surprise, instead of retreating, Rat No. 34 stubbornly scampered across his cage and pressed the bar over and over again. Rather than fearing the shocks as many other rats had done earlier, this rat hunted them down. The experimenters looked on as Rat No. 34 pushed the bar more than seven thousand times in twelve hours: once every five seconds without rest. Like an ultramarathon runner who deliriously refuses to stop for sustenance, the rat ignored a small trough of water and a tray of pellets. Sadly, he had eyes only for the bar. Twelve hours after the experiment began, Rat No. 34 was dead from exhaustion.

At first, Olds and Milner were confused. If every other rat avoided the shocks, why would Rat No. 34 do the opposite? Perhaps there was something wrong with his brain. Milner was ready to try the experiment with a different rat when Olds made a bold suggestion. Olds had once crawled around to imagine life as a fruit fly, and now he tried his hand at reading the mind of a rat. Considering Rat No. 34's behavior carefully, he became convinced that the rat was *enjoying* the shocks. It wasn't that he was seeking out pain, but rather that the shocks felt good. "The genius of Jim Olds was that he was open-minded enough and crazy enough to think that the animal *liked* being shocked," Aston-Jones said. "At the time, no one imagined that electrical stimulation in the brain could be pleasurable, but Olds was crazy enough to think the animal was having a good time."

So Olds investigated. He removed the probe from the rat's brain and noticed that it was bent. "Olds had been aiming for the mid-brain, but the probe bent into the rat's septum," says Aston-Jones. A fraction of an inch made all the difference between delight and discomfort. Olds took to calling this area of the brain the "pleasure center," a simplistic name that nonetheless captures the euphoria that rats—and dogs, goats, monkeys, and even people—feel when the area is stimulated. Some years later, when neuroscientist Robert Heath inserted an electrode into a depressed woman's pleasure center, she began to giggle. He asked why she was laughing, and though she couldn't offer an explanation, she told him that she felt happy for the first time in as long as she could remember. As soon as Heath removed the probe, the patient's smile disappeared. She was depressed again—and worse, she now knew what it felt like to be happy. She wanted more than anything for the probe to remain implanted, delivering regular shocks like a small hedonic pacemaker. Like Olds and Milner before him, Heath had shown how addictive euphoria could be.

———

After the demise of Rat No. 34, Olds and Milner found the same addictive behavior when they stimulated the pleasure center of other rats. Those rats, too, ignored food and water while

they pushed the little bar over and over again. Aryeh Routtenberg worked on some of these follow-up experiments, and he recalls that the rats behaved like addicts. The bar-pushing rats were
no different from rats that had addictive substances injected directly into their brains. "We threw all sorts of feel-good drugs at
the animals—amphetamines, chlorpromazine, monoamine oxi-
dase inhibitors—and they behaved just like the self-stimulating
rats." Routtenberg remembers an experiment that showed the
power of the pleasure center:

> One of the nice things about being a professor is that you can
> study whatever you like. I wanted to see what would happen
> if I made the bar-pressing animals drunk. I injected the alco-
> holic equivalent of a three-martini lunch into several rats,
> who just fell over. We lifted them up—as you'd drag a
> drunkard from the bar—and we led them over to the small
> metal bar. We laid them down so their heads brushed against
> the bar, which delivered a shock to their brains. In no time,
> these rats started pressing the bar over and over again. They
> were catatonic just a minute ago, but now they looked abso-
> lutely normal! After ten or fifteen minutes, we disabled the
> shocks, and the rats fell back into a stupor.

That wasn't the only reason why the researchers saw the rats
as tiny addicts. They showed the same restlessness that human
drug addicts show between hits. When the researchers prevented
the rats from shocking themselves more than once every few
minutes, the rats took to drinking lots of water to pass the time.
"The minute the reward stopped, they'd start drinking like

crazy," recalls Routtenberg. "I'd come back between experimental sessions and they were sitting there, completely bloated! It's
like they were doing something—anything—to pass the time.
The reward was so great that they would need to find a way to
pass the time until the next reward was available."

Word of the experiments got out, and the researchers began
to hear rumors. "We heard that the military was training goats,"
Bob Wurtz recalls. "They would guide the goats to bring am-
munition to soldiers, or even to carry bombs to the enemy." The
soldiers could encourage the goats to walk in a specific direction
by shocking or withholding shocks from the pleasure center. The
research influenced how experts like Wurtz, Aston-Jones, and
Routtenberg understood addiction. Olds and Milner originally
believed that Rat No. 34 was predisposed to be an addict. They
assumed that a problem with his internal wiring had driven him
to place electric stimulation above all else—even food, water, and
ultimately life. But at Olds' urging, they realized that there was
nothing wrong with Rat No. 34. He wasn't an addict by nature.
He was just an unfortunate rat that happened to be in the wrong
place at the wrong time.

———

This is one of the great lessons from Olds and Milner's experiment. Rat No. 34 behaved like an incurable addict but that
didn't mean there was something wrong with his brain. Like the
Vietnam G.I.s, he was a victim of circumstance. He was simply
responding as any rat would have done when a probe delivered
shocks to his pleasure center.

Routtenberg wondered if this might tell us something about

addiction in humans. But perhaps anyone could descend into
oblivion like Rat No. 34. "We started to think of addiction as a
form of learning. You can think of addiction as part of memory,"
says Routtenberg. Addicts had simply learned to link a particular
behavior with an appealing outcome. For Rat No. 34, this was
stimulation of his pleasure center; for a heroin addict, the flush of
pleasure from a fresh hit.

To measure the link between addiction and memory, Routtenberg visited the local pet store and bought a squirrel monkey
named Cleopatra. Ethics boards weren't as strict as they are now.
"I had my own lab room, so I could do whatever I wanted. I oper-
ated on her and put electrodes in the reward systems of her brain.
This had never been done before with a monkey." Routtenberg
placed Cleopatra in a cage in front of two metal bars. The first
sent an electrical current to her pleasure center, and the second
released a fresh supply of food. At first Cleopatra pushed the bars
randomly, but very quickly she began to behave like Rat No. 34,
ignoring the food bar and pressing the electric shock bar over
and over again. Olds saw what Routtenberg had done, and he
was delighted. "He came down to the lab with a friend, who was
a big-time researcher at Johns Hopkins, and showed him what
Cleopatra was doing," Routtenberg says. "It was one of the proud-
est days of my life." Later, Routtenberg removed Cleopatra from
the cage for hours or even days. Outside the cage, she detoxed,
becoming the same healthy monkey she had been when she
first arrived at the lab. But as soon as Routtenberg returned
her to the cage, she would frantically begin pressing the bar
again. Even when the bar was removed from the cage, she would

stand where it had once been. As Routtenberg guessed, Cleopatra's addiction had left a powerful imprint in her long-term
memory.

Jim Olds' lab held the solution to Lee Robins's conundrum.
The reason why her Vietnam vets escaped their heroin addictions was because they had escaped the circumstance that ensnared them. That was the case for Cleopatra, Aryeh Routtenberg's
squirrel monkey, who was every bit the addict inside her cage.
She pounded the metal bar that delivered shocks to her pleasure
center over and over again. She ignored her food and water. This
cage was to Cleopatra what Vietnam was to the bored G.I.s who
developed a taste for heroin. Cleopatra was healthy until she
joined the lab. When Routtenberg eventually removed her from
the cage, she became healthy again. But when she sat inside her
cage, the addiction returned with a vengeance.

Cleopatra returned to her cage, but few of the G.I.s ever
returned to Vietnam. They arrived home to a completely different life. There was no trace of the jungle; the steamy summers in
Saigon; the rattle of gunfire, or the chop of helicopter blades. Instead, they went grocery shopping, they returned to work, they
endured the monotony of suburbia, and enjoyed the pleasures of
home-cooked meals. Both Cleopatra and the soldiers showed that
Routtenberg was right: addiction embeds itself in memory. For
Cleopatra, the cage was a trigger. It transported her back to the
time when she had been an addict, and she couldn't help falling
back on old habits. The lucky Vietnam vets never confronted

those memories, because once they left Vietnam they escaped the cues that went along with the act of shooting up.

This is why most heroin users struggle to stay clean. Like Cleopatra, they return to the scene of the crime over and over again. They see friends who remind them of a time when they were addicts; they live in the same homes; they walk through the same neighborhoods. Nothing changes once they're clean, except the fact that instead of giving in to the addiction, they're resisting it every day. This is why the temptation is so great. What else are they supposed to do when every sight, smell, and sound rekindles the moment of bliss that follows a hit?

***

Isaac Vaisberg, a former gaming addict, knows the dangers of returning to the scene of the crime. Nothing marks Isaac as a natural candidate for addiction. He was born in Venezuela in 1992 to a wonderfully supportive mom and an overworked but attentive dad. When Isaac was a boy, his parents divorced and he moved to Miami with his mom. His dad remained in Venezuela, but the two talked often, and Isaac visited when he wasn't in school. His grades were stellar, rarely dropping below an *A*. At the end of his junior year of high school, he scored 2200 out of a possible 2400 on the SATs, which placed him in the top 1 percent of all students in the United States. He was admitted to Worcester Academy, one of the country's most competitive boarding schools, not too far from Boston, and later to American University in Washington, D.C. Isaac wasn't just a scholar—he was also an athlete. Worcester granted him a football scholarship, and he arrived in great physical shape, ready to play as a first-string linebacker.

Unfortunately, that's only half the story. Isaac was lonely. "My parents got divorced when I was very young, and I ended up ping-ponging between the United States and Venezuela. Because of that ping-ponging, I was adept at forming new relationships, but not very good at forming deep relationships." Instead, he found friends online.

When Isaac was fourteen, he started playing World of Warcraft. WoW is addictive for a number of reasons, but Isaac found the game's social dimension irresistible. Like many players, he joined a guild, a small band of players who share resources and chat regularly in guild-specific chat rooms. His guild-mates became his closest friends, and their friendship ultimately came to stand in for the meaningful relationships he lacked in the off-line world.

Isaac's first dangerous binge began during his junior year in high school. "I had picked up and dropped World of Warcraft many times, but this time it became my sole means of socializing and my sole release. I got a small dopamine hit every night, and it helped me overcome my anxieties." He stopped sleeping, his grades plummeted, and he became physically sick when his mother insisted he go to school. "I would flip out and have panic attacks. Getting in the car in the morning I'd feel nauseated. The second I knew I didn't have to go to school, these symptoms went away." Isaac ultimately recovered after this first binge, and by the end of that school year he was doing so well that he aced his SATs.

Isaac's second binge began a couple of months into his time at Worcester Academy. Left in his dorm room without supervision, he rejoined his old guild and rekindled the online friendships he'd formed the year before. Soon it became an obsession again. "When I arrived at Worcester Academy, I weighed about one-ninety-five. I was fit and playing football. By the end of the first semester, I weighed about two-thirty-five. I lost a significant amount of hair on my head, quit the football team, and had *C*s across the board." Isaac was resilient, though. He managed to complete his senior year and gain acceptance to American University. At this point he still believed his binges were flukes. He wasn't concerned that his addiction might follow him to college.

His first semester at American was a success—he aced his classes and remained fit and healthy. In his second semester, though, he became stressed. He decided to "play just a bit" of WoW as a release, and ultimately failed his second semester classes. Isaac's transcript was a roller coaster of *A*s and *F*s, and his mom was so worried that she arrived unannounced and presented him with a pamphlet for the reSTART addiction recovery center, located just outside Seattle. He agreed to enroll in an in-patient program, but only after logging in to his WoW account to tell his guild-mates that he'd be offline for a while.

reSTART is the world's first gaming and Internet addiction treatment center. Its founders recognize that Internet use differs from substance addiction, because it's almost impossible to return to society without using the Internet. You can hold a job, pay your bills, and communicate without using drugs and alcohol, but not without using the Internet. Echoing the green movement, the center therefore aims to teach patients how to use the

Internet "sustainably," rather than encouraging them to avoid it altogether.

Isaac began his six-week program with enthusiasm, making friends, painting, hiking the spectacular trails around the center, and regaining his strength at its gym. He formed close bonds with some of the mentors, who told him that WoW had given him an illusion of control over his life. Outside the game his world had continued to crumble, but that seemed to matter less and less as he conquered one WoW quest after another. Despite making good progress, at times he felt frustrated. Though reSTART had helped, Isaac saw his time there as a roadblock that distracted him from finishing college and moving on to a healthier, self-sufficient phase of his life. He couldn't really be "better" until he settled back into the real world. Though he went as far as buying an airline ticket to D.C. online, he ultimately stayed for the full six weeks.

Then Isaac made his biggest mistake. "I got through the rest of the program, my chest puffed up, and I was a little bit more confident in what I was doing. But when it came time to present my life balance plan at the end of the program, the one thing everybody criticized was my decision to go back to D.C." Isaac describes this using the language you'd expect from a veteran gamer: "I just felt like I couldn't leave something unconquered. I couldn't leave American University without my degree—it just wasn't gonna happen. Against medical advice I decided to go back east."

Isaac's experience differs from the lives of Lee Robins's Vietnam vets. Instead of escaping the context of his addiction forever, Isaac returned to D.C. For two or three months, things went well.

He got a job, he started working as a math tutor and made good money, and his guidance counselor admitted him back into American University. Things were looking up—until they weren't.

Isaac told me that the most dangerous time for an addict is the first moment when things are going so well that you believe you've left the addiction behind forever. "You're convinced that you're fixed, so you can go back to doing what you were doing before. I let my guard down, and a buddy of mine sent me a text message that said, 'Hey, you wanna play with us a little bit?' And I went, 'Hey, sure!'"

That was Thursday, February 21, 2013. Isaac is sure of the date, because it left an indelible imprint in his memory. Two days later he was scheduled to tutor a kid who had an algebra exam, but he missed the appointment. He didn't go to class on Monday either, and then he spent five weeks alone in his apartment. He didn't leave once and he didn't shower. In exchange for a small tip, his doorman brought food he ordered by phone to his room. His place began to smell and empty containers towered around his desk. He played twenty hours a day and collapsed, numb, for a few hours of sleep before returning to the game when he awoke. He completed one mission after another, chatted with his guild-mates for days, and lost touch with the outside world. Five weeks passed quickly. He missed one hundred and forty-two phone calls (another number he says he can't forget), but, for some reason that escapes him even now, he decided to answer the one hundred and forty-third call. It was his mother, and she told him that she was visiting in two days.

After one final binge, he decided to clean his apartment and take a shower. This was his "rock-bottom moment." He was disgusted by what he saw in the mirror. He'd put on sixty pounds of pure fat, his hair was greasy, and his clothes were filthy. He described a recurring vision that, even eighteen months later, brought him close to tears:

When I was growing up, my dad didn't have a lot of money. He started a business, and left for work at five in the morning and came home around nine at night. He was very happy when he got home. He'd give me a huge hug, grab a little glass of Scotch, go to his chair by the window and open it up so he could enjoy the breeze. And then he'd do it all over again, every single day.

I had this image of him walking into my apartment, and grabbing a little glass of Scotch, going to his chair, and crying. I had never seen my dad cry. He always had his chest up, and he was always strong. And I imagined him crying in his chair, wondering what he did wrong with me. It hurts just talking about it. It was this burning pain in my heart that he would feel that way for *my* fuck-up.

Isaac took his mom to dinner, where he broke down and told her he'd fallen off the wagon. He told her he needed to try reSTART again, but this time with a better attitude. He wouldn't return to D.C., and after the six-week inpatient program ended, he'd enroll in a seven-month after-care outpatient program.

Isaac was true to his word. He embraced the inpatient program and felt comforted knowing that the outpatient program would give him extra support as he grew used to living and working outside the center. The outpatient program made all the difference. Like other outpatients, Isaac spent between twenty and thirty hours at the center each week, while also holding down a part-time job. He lived with several former inpatients, who supported each other and vigilantly ensured that their roommates didn't relapse.

Isaac decided to stay in the Seattle area, near reSTART. He visits the center often, but now spends most of his time running a CrossFit gym. In April 2015, he bought the gym from its former owners, and after just four months under his care its membership tripled. The gym gives him a healthy way to fulfill his psychological needs: he has plenty of friends, remains active and healthy, and sets business-oriented goals that keep him motivated.

Isaac Vaisberg, like Robins, Milner, Olds, and their students, taught the world a profound lesson about addiction and its victims: there's so much more to addiction than an *addictive personality*. Addicts aren't simply weaker specimens than non-addicts; they aren't morally corrupt where non-addicts are virtuous. Instead, many, if not most, of them are unlucky. Location isn't the only factor that influences your chances of becoming an addict, but it plays a much bigger role than scientists once thought. Genetics and biology matter as well, but we've recognized their role for decades. What's new, and what only became clear in the 1960s and 1970s, is that addiction is a matter of environment, too. Even the sturdiest of our ranks—the young G.I.s who were free of addiction when they left for Vietnam—are prone to weakness when they find themselves in the wrong setting. And even the most determined addicts-in-recovery will relapse when they revisit the people and places that remind them of the drug.

Time has made a fool of the experts who once believed that addiction was reserved for a wretched minority, because, like Isaac Vaisberg, tens of millions of people in the developed world today exhibit one or more behavioral addictions. The very concept was foreign to Olds and Milner in the 1950s, and to Robins in the 1970s. People were addicted to substances—not behaviors. The feedback they got from behaviors alone could never rise to the euphoric intensity of injected heroin. But just as drugs have become more powerful over time, so has the thrill of behavioral feedback. Product designers are smarter than ever. They know how to push our buttons and how to encourage us to use their products not just once but over and over. Workplaces dangle carrots that always seem to be just out of reach. The next promotion is around the corner; the next sales bonus is one sale away.

As for Rat No. 34, hammering away at the bar in his cage, our brains host a flurry of electrical activity when we're engaged with an addictive behavior. For decades, researchers believed this activity was the root of addiction: mimic the right brain patterns and you'd create an addict. But the biology of addiction is far more complicated than simply stimulating a clump of neurons. Addiction, as it was for Isaac Vaisberg, the Vietnam vets, and Rat No. 34, is a matter of learning that the addictive cue—a game, a place paired with heroin, or a small metal bar—treats loneliness, disaffection, and distress.

3.

The Biology of
Behavioral Addiction

There's a modern-day malady that affects two thirds of all adults. Its symptoms include: heart disease, lung disease, kidney disease, appetite suppression, poor weight control, weakened immune functioning, lowered resistance to disease, higher pain sensitivity, slowed reaction times, mood fluctuations, depressed brain functioning, depression, obesity, diabetes, and certain forms of cancer.

That malady is chronic sleep deprivation, which is rising in the wake of smartphones, e-readers, and other light-emitting devices. Sleep deprivation is behavioral addiction's partner—the consequence of persistent overengagement. It's a global problem that has recently attracted plenty of attention, including from entrepreneur and author Arianna Huffington. At the 2016 World Economic Forum in Davos, Huffington discussed her forthcoming book on sleep, titled *The Sleep Revolution*:

I got an email two hours ago from the official Davos establishment, which was a sleep survey of the world. It shows
that people spend more time on their digital devices than
sleeping . . . I think it's really interesting to look at the relationship between technology and taking care of ourselves.
Because we're obviously all addicted to technology. So how
can we put it in its place? And not on your nightstand. That
is the key guys—do not charge your phones by your bed.

Huffington was wise to focus on smartphone charging.
Ninety-five percent of adults use an electronic device that emits
light in the hour before bed, and more than half check their
emails overnight. Sixty percent of adults aged between eighteen
and sixty-four keep their phones next to them when they sleep,
which might explain why 50 percent of adults claim they don't
sleep well because they're always connected to technology. Sleep
quality has declined dramatically in the past half century, particularly over the past two decades, and one of the major culprits
is the bluish light that emanates from many of these electronic
devices.

For millennia, blue light existed only during the daytime.
Candles and wood fires produced reddish-yellow light, and there
was no artificial lighting at night. Firelight isn't a problem, because the brain interprets red light as a signal for bedtime. Blue
light is a different story, because it signals morning. So 95 percent
of us are inducing jet lag at night by telling our bodies that the
day is beginning just before we go to bed.

Normally, the pineal gland buried deep in your brain produces a hormone called melatonin at night. Melatonin makes you sleepy, which is why people who suffer jet lag take melatonin supplements before bed. When blue light hits the back of your eyes, the pineal gland stops producing melatonin, and your body prepares for the day. In 2013, a group of scientists measured how much melatonin thirteen volunteers produced after using an iPad for two hours late at night. When those volunteers wore orange goggles—to simulate evening light—they produced plenty of melatonin, which prepared their bodies for bed. When they wore blue goggles (and to some extent when they used the iPad without goggles), their bodies produced significantly less melatonin. The researchers urged "manufacturers to design [sleep-cycle]-friendly electronic devices" with backlights that turned progressively more orange at night. A second study, this time without goggles, found the same effect: people produce less melatonin, sleep more poorly, and feel more tired when they use an iPad before bed. In the long run, our technology compulsions are damaging our health.

As much as blue light hampers our ability to sleep, the real damage of behavioral addiction happens when we're wide awake, obsessively juggling laptops and tablets, fitness trackers and smartphones.

The human brain exhibits different patterns of activity for different experiences. One clump of neurons fires when you imagine your mother's face; a different clump when you imagine the house where you grew up. These patterns are fuzzy, but by looking at a person's brain you can tell roughly whether she's thinking about her mother or her first home.

There's also a pattern that describes the brain of a drug addict as he injects heroin, and a second that describes the brain of a gaming addict as he fires up a new World of Warcraft quest. They turn out to be almost identical. Heroin acts more directly, generating a stronger response than gaming, but the patterns of neurons firing across the brain are almost identical. "Drugs and addictive behaviors activate the same reward center in the brain," according to Claire Gillan, a neuroscientist who studies obsessive and repetitive behaviors. "As long as a behavior is rewarding—if it's been paired with rewarding outcomes in the past—the brain will treat it the same way it treats a drug." What makes drugs like heroin and cocaine more dangerous in the short-term is that they stimulate the reward center much more strongly than behaviors do. "Cocaine has more direct effects on the neurotransmitters in your brain than, for example, gambling, but they work by the same mechanism on the same systems. The difference is in their magnitude and intensity."

This idea is quite new. For decades, neuroscientists believed that only drugs and alcohol could stimulate addiction, while people responded differently to behaviors. Behaviors might be pleasurable, they suggested, but that pleasure could never rise to the destructive urgency associated with drug and alcohol abuse. But more recent research has shown that addictive behaviors produce the same brain responses that follow drug abuse. In both cases, several regions deep inside the brain release a chemical called dopamine, which attaches itself to receptors throughout the brain that in turn produce an intense flush of pleasure. Most of the time the brain releases only a small dose of dopamine, but certain substances and addictive experiences send dopamine production into overdrive. Warming your hands by a log fire on a cold night or taking a sip of water when you're thirsty feels good, but that sensation is dramatically more intense for an addict when he injects heroin or, to a lesser extent, begins a new World of Warcraft quest.

At first the upsides dramatically outweigh the downsides as the brain translates the rush of dopamine into pleasure. But soon the brain interprets this flooding as an error, producing less and less dopamine. The only way to match the original high is to up the dosage of the drug or the experience—to gamble with more money or snort more cocaine or spend more time playing a more involving video game. As the brain develops a tolerance, its dopamine-producing regions go into retreat, and the lows between each high dip lower. Instead of producing the healthy measure of dopamine that once inspired optimism and contentment in response to small pleasures, these regions lie dormant until they're overstimulated again. Addictions are so pleasurable that the brain does two things: first it produces less dopamine to dam the flood of euphoria, and then, when the source of that euphoria vanishes, it struggles to cope with the fact it's now producing far less dopamine than it used to. And so the cycle continues as the addict seeks out the source of his addiction, and the brain responds by producing less and less dopamine after each hit.

———

As a kid I was terrified of drugs. I had a recurring nightmare that someone would force me to take heroin and that I'd become addicted. I knew very little about addiction, but I pictured myself frothing at the mouth in a bleak treatment center. As time passed I realized that drug pushers weren't going to waste their time on a neurotic seven-year-old, but one part of the nightmare stuck with me: the idea that a person could become addicted against his will; that if you happened to come into contact with an addictive substance, you'd develop an addiction. If addiction were simply a brain disorder, my seven-year-old self would have been right: flood the brain with dopamine and you create an addict. But that's not how addiction works at all. Since your brain fundamentally reacts the same way to any pleasurable event, there has to be another ingredient—otherwise we'd all develop crippling ice cream addictions from an early age. (Just imagine the dopamine shock that follows a toddler's first taste of ice cream.)

The missing ingredient is the situation that surrounds that rise in dopamine. The substance or behavior itself isn't addictive until we learn to use it as a salve for our psychological troubles. If you're anxious or depressed, for example, you might learn that heroin, food, or gambling lessen your pain. If you're lonely, you might turn to an immersive video game that encourages you to build new social networks.

"We have systems for parenting and love, and those systems push us to persist despite negative consequences," Maia Szalavitz, a writer who focuses on addiction, explains. "The system that's designed for that sort of behavior is the template for addiction. When this system becomes misaligned, you get addictions." Each of the systems that Szalavitz refers to is a collection of instinctive survival behaviors, like the drive to care for your children or to find a romantic partner. The same instincts that push us to persevere in the face of pain and difficulty can also propel fanaticism and damaging addictive behavior.

In one article, Szalavitz explains that no one else can turn you into an addict. "Pain patients cannot be 'made addicted' by their doctors," Szalavitz says. "In order to develop an addiction, you have to repeatedly take the drug for emotional relief to the point where it feels as though you can't live without it . . . it can only happen when you start taking doses early or take extra when you feel a need to deal with issues other than pain. Until your brain learns that the drug is critical to your emotional stability, addiction cannot be established." Addiction isn't just a physical response; it's how you respond to that physical experience psychologically. To underscore the point, Szalavitz turns to heroin, the most addictive and dangerous illicit drug. "To put it bluntly, if I kidnap you, tie you down, and shoot you up with heroin for two months, I can create physical dependence and withdrawal symptoms—but only if you go out and use after I free you will you actually become an addict."

"Addiction isn't about 'breaking' your brain, or 'hijacking' your brain, or 'damaging' your brain," Szalavitz says. "People can be addicted to behaviors, and even to the experience of love. Addiction is really about the relationship between the person and the experience." It isn't enough to ply someone with a drug or a behavior—that person also has to learn that the experience is a viable treatment for whatever ails them psychologically.

The highest risk period for addiction is early adulthood. Very few people develop addictions later in life if they haven't been addicted in adolescence. One of the major reasons is that young adults are bombarded by a galaxy of responsibilities that they're not equipped to handle. They learn to medicate by taking up substances or behaviors that dull the insistent sting of those persistent hardships. By their midtwenties, many people acquire the coping skills and social networks that they lack in adolescence. "If you aren't using drugs as a teenager, you're probably also learning to deal with your troubles using other methods," Szalavitz said. So you develop a degree of resilience by the time you emerge through the gauntlet of adolescence.

The most striking thing Szalavitz told me was that addiction is a sort of misguided love. It's love with the obsession but not the emotional support. That idea might sound fluffy, but it's grounded in science.

In 2005, an anthropologist named Helen Fisher and her colleagues placed infatuated lovers in a brain scanner. She described their findings in an article titled "Love Is Like Cocaine":

> I felt like jumping in the sky. Before my eyes were scans showing blobs of activity in the ventral tegmental area, or VTA, a tiny factory near the base of the brain that makes dopamine and sends this natural stimulant to many brain regions . . . This factory is part of the brain's reward system, the brain network that generates wanting, seeking, craving, energy, focus, and motivation. No wonder lovers can stay awake all night talking and caressing. No wonder they become so absent-minded, so giddy, so optimistic, so gregarious, so full of life. They are high on natural "speed." . . . Moreover, when my colleagues re-did this brain scanning

experiment in China, their Chinese participants showed just as much activity in the VTA and other dopamine pathways— the neurochemical pathways for wanting. Almost everyone on earth feels this passion.

In the 1970s, a psychologist named Stanton Peele published *Love and Addiction*, explaining that the very healthy attachment we feel toward people we love can also be destructive. This same attachment could be directed toward a bottle of vodka, a syringe of heroin, or an evening at the casino. They're impostors because they soothe psychological discomfort in the same way that social support makes hardship easier—but they soon replace short-term pleasure with protracted pain. The capacity for love is the result of millennia of evolution. This makes people well-designed to raise offspring and to shepherd their genes into the next generation—but also susceptible to addiction.

Destructiveness is a critical part of addiction. There are many ways to define addiction, but the broadest definitions go too far because they include acts that are healthy or essential for survival. In a 1990 editorial in the *British Journal of Addiction*, a psychiatrist named Isaac Marks claimed that, "Life is a series of addictions and without them we die." Marks titled the editorial "Behavioral (Non-Chemical) Addictions," and he was being provocative for good reason. Behavioral addictions were relatively new to the field of psychiatry:

> Every few moments we inhale air. If deprived of it, within
> seconds we strive to breathe, with immense relief when we
> succeed. More prolonged deprivation causes escalating ten-

sion, severe withdrawal symptoms of asphyxiation and death within minutes. On a longer time scale, eating, drinking, defaecation, micturation and sex also involve rising desires to perform an act; the act switches off the desire, which returns within hours or days.

Marks was right: breathing seemed to mirror the properties of other addictions. But the idea of addiction isn't interesting or useful if it describes every single activity that plays a role in our survival. It doesn't make sense to call a cancer patient an addict because she needs her chemotherapy medication. Addictions should, at the very least, leave our chances of surviving unchanged; as soon as they mirror the life-sustaining properties of breathing, eating, and chemotherapy drugs, they're no longer "addictions."

Stanton Peele linked love and addiction in the 1970s, arguing that love drove addiction when it was misdirected and turned toward dangerous targets. Like Marks fifteen years later, Peele was also arguing that addiction went beyond illegal drugs. That had been the position of scientists for decades, so much so that few of them were willing to accept that nicotine was addictive. Since smoking was legal, by their logic, its component parts couldn't possibly be addictive. The term "addiction" had become so stigmatized that it was reserved for a small, closed set of substances. But the term wasn't sacred to Peele. He pointed out that many smokers leaned on nicotine in the same way that heroin addicts relied on heroin as a psychological crutch, although heroin was more obviously damaging in the short-term. Peele's perspective was heretical in the 1970s, but the medical world caught

up in the 1980s and 1990s. Peele also recognized that any destructive crutch could become a source of addiction. A bored
white-collar worker who turned to gambling for the thrill he
lacked in the real world could develop a gambling addiction.

I approached Peele in researching this book, but he bristled
when I mentioned behavioral addiction. "Sure," he told me, signaling that he'd be happy to talk, "except I've never in my life
used the term 'behavioral addiction.'" To Peele the term was
heretical, because it implied there was a meaningful difference
between behavioral and substance addictions, a distinction he
argues doesn't exist because addiction isn't about substances or
behaviors or brain responses. Addiction, to Peele, is "an extreme,
dysfunctional attachment to an experience that is acutely harmful
to a person, but that is an essential part of the person's ecology
and that the person cannot relinquish." That's how he defined it
decades ago, and that's how he sees it today. The "experience" is
everything about the context: the anticipation of the event, and
the behavior of carefully lining up the needle, the charred spoon,
and the lighter. Even heroin—an addictive substance if ever there
was one—makes its way to the body via a chain of behaviors that
themselves become part of the addiction. If even heroin addiction
is to some extent "behavioral," you can see why Peele avoided the
term altogether.

Peele may not have used the term "behavioral addiction," but
for decades he has separated addictive behaviors and addictive
substances in his books. For example, the sixth chapter of Peele's
book, *The Truth About Addiction and Recovery*, written with psychiatrist Archie Brodsky in 1991, is titled "Addictions to Gam-

bling, Shopping, and Exercise." Peele and Brodsky asked, "Can one be addicted to gambling, shopping, exercise, sex, or love in the same sense that one is addicted to alcohol or drugs?" Their answer was *yes*—that "any activity, involvement, or sensation that a person finds sufficiently consuming can become an addiction . . . addiction can be understood only in terms of the overall experience it produces for a person . . . and how these fit in with the person's life situation and needs." Peele and Brodsky were also quick to dismiss the idea that any pleasurable, endorphin-producing activity was an addiction. "Endorphins don't make people run until their feet bleed or eat until they puke," they argued. Just because runners experience a "high" doesn't make them addicts. They refused to call gambling, shopping, and exercise compulsions "diseases," but they allowed that those activities were capable of inspiring addictive behaviors.

Peele was marginalized for decades. He railed against abstinence and Alcoholics Anonymous, and wrote again and again that addiction wasn't a disease. Rather, it was the association between an unfulfilled psychological need and a set of actions that assuaged that need in the short-term, but was ultimately harmful in the long-term. Peele was often inflammatory and always provocative, but his central message was unchanged: that any experience could be addictive if it seemed to soothe psychological distress. Peele's ideas have slowly drifted to the mainstream. Though the American Psychiatric Association (APA) still considers addiction a disease, four decades after Peele first linked love and addiction, the APA has acknowledged that addiction isn't limited to substance abuse.

Every fifteen years or so the APA releases a new edition of its bible, the *Diagnostic and Statistical Manual of Mental Disorders* (DSM). The DSM catalogs the signs and symptoms of dozens of psychiatric disorders, from depression and anxiety to schizophrenia and panic attacks. When the APA released the fifth edition of the DSM in 2013, it added *behavioral addiction* to its list of official diagnoses, and abandoned the phrase *substance abuse and dependence* in favor of *addictions and related disorders*. Psychiatrists had been treating behavioral addicts for years, and now the APA was catching up.

The APA also made clear that merely depending on a substance or behavior wasn't enough to warrant a diagnosis of addiction. Many hospital patients depend on opiates, for example, but that doesn't make all hospital patients opium addicts. The missing ingredients are the sense of craving that comes from an addiction, and the fact that addicts know they're ultimately undermining their long-term well-being. A hospital patient who relies on morphine while he recovers from surgery is doing what's best both in the short-term and the long-term; a morphine addict knows that his addiction combines short-term bliss and long-term damage. A number of current and former behavioral addicts told me the same thing: that consummating their addictions is always bittersweet. It's impossible to forget that they're compromising their well-being even as they enjoy that first rush of gratification.

The APA is only now endorsing the link between substance addiction and behavioral addiction, but isolated researchers have been making similar claims for decades. In the 1960s, even before Peele began publishing his ideas, a Swedish psychiatrist named Gösta Rylander noticed that dozens of tormented drug addicts were behaving like distressed wild animals. When confined to small spaces, animals soothe themselves by repeating the same actions over and over again. Dolphins and whales swim in circles, birds pluck their own feathers, and bears and lions pace within their enclosures for hours. By some reports, 40 percent of caged elephants march in circles and rock back and forth in a desperate quest for comfort.

These are universal signs of distress, so Rylander was worried to see similar behavior in regular amphetamine users. One patient collected and arranged hundreds of rocks by size and shape, and then jumbled them so he could begin the process from scratch. Dozens of motorcyclists in a gang of amphetamine users rode around the same suburban block two hundred times. A man picked at his hair incessantly, and a woman filed her nails for three days until they bled. When Rylander asked them to explain what they were doing, they struggled to concoct sensible answers. They knew they were behaving strangely, but they felt compelled to continue. Some of them were driven by an intense pathological curiosity, while others found the act of repetition soothing. Rylander reported what he saw in a journal article, where he labeled the behavior *punding*, a Swedish word that means blockheadedness or idiocy. Most interesting to Rylander, though, was that for these patients there was no line between drug addiction and behavioral addiction. One bled into the other, and they were similarly harmful, soothing, and irresistible.

Rylander died in 1979, but left a significant legacy. A growing circle of doctors and researchers reported punding in cocaine addicts and other drug users, and Rylander's paper was cited hundreds of times. Punding behaviors are often bizarre, but they affected exactly who experts might have predicted: heavy drug users. That was true, at least, until the early 2000s, when a small group of neuroscientists began to see punding and other odd repetitive behaviors in the least likely of suspects.

———

In the early 2000s, Andrew Lawrence, a neuroscience professor at Cardiff University, and some of his colleagues noticed a range of strange addictive behaviors in people suffering from Parkinson's disease. There's almost no overlap between the stereotypical personalities of heavy drug users and Parkinson's patients. Where drug users are young and impulsive, Parkinson's patients tend to be elderly and sedate. More than anything, they hope to enjoy the final decades of their lives without suffering through the muscle tremors that are typical of the disease. The only overlap, in fact, is that these Parkinson's patients were using a very strong drug to treat their tremors. "Parkinson's results from a dopamine deficit, so we treat the disease with drugs that replace dopamine," Lawrence said. Dopamine is produced by a number of brain regions, and it produces a wide variety of effects. It controls motion (hence the tremors in Parkinson's patients) and plays a major role in shaping how people respond to rewards and pleasure. Dopamine targets Parkinsonian tremors, but also happens to introduce a form of pleasure or reward. Many patients, left to their own devices, develop addictions to dopamine replacement drugs, so neurologists monitor their dosages very closely. But that wasn't what fascinated and troubled Lawrence most.

Patients were hoarding their medication, and we happened to notice that some of them were also displaying behavioral addictions," Lawrence said. "So they would report problem gambling, problem shopping, binge-eating, and hypersexuality." In 2004, Lawrence catalogued some of these symptoms in a staggering review paper. One man, an accountant who had been a dedicated and careful saver for half a century, developed a gambling habit. He had never gambled before, but suddenly he felt drawn to the thrill of risk. At first he gambled conservatively, but soon he was gambling a couple of times a week, and then every day. His hard-won retirement savings shrank slowly at first, and then more quickly, until he went into debt. The man's wife panicked and asked their son for money, but their son's contribution merely fueled the man's addiction. One day his wife found the man rummaging through the garbage, hoping to retrieve the lottery tickets she'd torn up earlier that day. Worst of all, the man couldn't explain the change in his character. He didn't want to gamble, or to squander his life savings, but he couldn't help himself. When he fought the tendency to gamble it occupied his every thought. Only gambling seemed to relax him.

Other elderly patients developed sexual fetishes, and pestered their husbands and wives for sex throughout the day. One man, a lifelong fashion conformist, took to dressing up like a prostitute. Others developed addictions to Internet pornography. Lifelong health nuts binged on candy and chocolates and put on mountains of weight in a few short months. Strangest of all, perhaps, was the man who couldn't stop giving away his money. When his bank account was empty, he began giving away his possessions instead. When Billy Connolly, the celebrated Scottish comedian,

developed Parkinson's in his late sixties, he began taking dopamine replacement drugs. He, too, succumbed to behavioral addictions and had to stop treatment. "The doctors took me off the
medication, because the side effects were stronger than the effects," Connolly told Conan O'Brien on a late-night talk show
appearance. "I asked what the side effects were, and they said, 'an
overriding interest in sex and gambling.'" Connolly makes light
of the anecdote on TV, but without treatment his tremors are
becoming increasingly severe. The drugs are so strong that up to
half of all patients seem to develop some of these side effects.

Lawrence argued that these patients were simply enacting
whatever behaviors came to them most naturally. These behaviors, called *stereotypies*, depend on "individual life histories,"
Lawrence wrote. "For example, office workers stereotypically
shuffle papers, a seamstress will collect and arrange buttons." A
sixty-five-year-old businessman repeatedly dismantled and re-
constructed pens, and tidied an already immaculate space on his
desk. A fifty-eight-year-old architect tore down and reconfigured
his home office over and over again. A fifty-year-old carpenter
collected hardware tools and unnecessarily felled a tree in his
yard. These familiar actions became a source of comfort because
they came so fluently and demanded very little thought.

Lawrence and Rylander before him were witnessing the
blurred line between substance addictions and behavioral addictions. Like drugs or alcohol, stereotypies offered just one more
route to soothe a tormented psyche. Lawrence pointed out this
overlap by noting that many of the patients who were stuck in a
behavior loop also overdosed on their dopamine-producing medication. Those with aggressive Parkinson's were often fitted with

a small pump that administered the drugs internally. Though they were told to obey a schedule, they could push a button to administer a fresh dose of the drug when their symptoms flared. Many of them began by following the schedule, but they soon learned that the drug also made them feel good. Some of the patients who became addicted to the drug also developed behavioral addictions, and they would jump back and forth between the two. One day they might take a few extra doses of the drug, and the next they might shuffle papers for several hours in the morning before collecting and arranging rocks from the garden in the afternoon. Sometimes they'd do both at the same time, self-medicating with both drugs and soothing behaviors. There was no material difference between these two routes to addiction; they were essentially two versions of the same malfunctioning program.

I n the 1990s, a neuroscientist at the University of Michigan named Kent Berridge was trying to understand why addicts continued using drugs as their lives deteriorated. One obvious answer was that addicts get so much pleasure from their addictions that they're willing to sacrifice long-term well-being for a jolt of immediate bliss—that they fall in dysfunctional love with a partner that destroys them in return. "Twenty years ago we were looking for mechanisms of pleasure," Berridge said. "And dopamine was the best mechanism of pleasure out there, and everybody knew it was involved in addiction. So we set out to gather more evidence to show that dopamine was a mechanism of pleasure." To Berridge and many other researchers the link seemed obvious—so obvious that he expected to find it quickly so he could move on to answer newer, more interesting questions.

But the result turned out to be elusive. In one experiment, Berridge gave rats a delicious sugary liquid and watched as they licked their lips with pleasure. "Like human infants, rats lick their lips rhythmically when they taste sweetness," Berridge said. Rat researchers learn to interpret different rattish expressions, and this one was the gold standard for pleasure. Based on his understanding of dopamine, Berridge assumed that each rat's tiny brain was flooding its host with dopamine each time it tasted the sweet liquid, and this rise in dopamine drove the rat to lick its lips. Logically, if Berridge stopped the rat from producing dopamine, it should stop licking its lips. So Berridge performed a kind of brain surgery on the rats to stop them from producing dopamine, and fed them the liquid again.

The rats did two things after surgery, one of which surprised Berridge and one of which didn't. As he expected, they stopped drinking the sugary liquid. The surgery had knocked out their appetite by preventing their brains from producing dopamine. But the rats continued to lick their lips when he fed them the sugar water directly. They didn't seem to want it—but when they tasted it, they seemed to get just as much pleasure as they had before the surgery. Without dopamine they lost their appetite for sugar water, but still enjoyed it when they tasted it anyway.

"It took about ten years for this to sink in among the neuroscience community," Berridge says. The findings contradicted what neuroscientists long felt they knew to be true. "For a number of years people in the neuroscience world told us, 'No, we know dopamine drives pleasure; you have to be wrong.' But then evidence started to come in from studies on humans, and now very few researchers doubt our findings. In those studies, researchers would give people cocaine or heroin, as well as a second drug that was designed to block dopamine production. Blocking dopamine didn't reduce the pleasure they felt—but it did reduce the amount they took."

Berridge and his colleagues had shown that there was a big difference between liking a drug and wanting a drug. Addiction was about more than just liking. Addicts weren't people who happened to like the drugs they were taking—they were people who *wanted* those drugs very badly even as they grew to dislike them for destroying their lives. What makes addiction so difficult to treat is that wanting is much harder to defeat than liking. "When people make decisions, they privilege wanting over liking," Berridge said. "Wanting is much more robust and big and broad and powerful. Liking is anatomically tiny and fragile—it's easily disrupted and it occupies only a very small part of the brain. In contrast, it's not easy to disrupt the activation of an intense want. Once people want a drug, it's nearly permanent— it lasts at least a year in most people, and may last almost a whole lifetime." Berridge's ideas explain why relapse is so common. Even after you come to hate a drug for ruining your life, your brain continues to want the drug. It remembers that the drug soothed a psychological need in the past, and so the craving remains. The same is true of behaviors: even as you come to loathe Facebook or Instagram for consuming too much of your time, you continue to want updates as much as you did when they still made you happy. One recent study suggests that playing hard to get has the same effect: an unattainable romantic partner is less likable but more desirable, which explains why some people find emotionally unavailable partners alluring.

Liking and wanting overlap most of the time, which clouds their differences. We tend to want things that we like, and vice versa, because most pleasant things are good for us, and most unpleasant things are bad for us. The baby rats in Berridge's studies had evolved to instinctively like the taste of sugar water, because sweet substances tend to be both harmless and rich in calories. Their ancestor rats who gravitated toward sweet foods tended to live longer and to mate with other rats, so their sweet-tooth proclivity was passed down from one generation to the next. The rats who ate bitter foods were more likely to die, either from poisoning or from malnutrition. Very few truly bitter foods are packed with nutrients, and from a young age we avoid the many bitter plants and roots that happen to be toxic. Though they're often linked, Berridge showed that liking and wanting take different paths in the case of addiction. The depths of addiction are no fun at all, which is another way of saying that addicts crave a hit without *liking* the experience. Stanton Peele likened addiction to misguided love, and falling in love with the wrong person is a classic case of wanting without liking. Loving the wrong person is so common that we have stereotypes for the "guy who's no good" and the "femme fatale." We know they're no good for us, but we can't help wanting them.

Although Berridge spends more time investigating drug addiction, like Stanton Peele and Andrew Lawrence he believes his ideas also apply to behavioral addictions. "We always knew drugs could influence these brain systems, but we didn't know the same about behaviors. Over the past fifteen years or so, we've come to learn that the same is true of behaviors—and the process works through the same brain mechanisms." Just as drugs trigger dopamine production, so do behavioral cues. When a gaming addict fires up his laptop, his dopamine levels spike; when an exercise addict laces her running shoes, her dopamine levels spike. From there, these behavioral addicts look a lot like drug addicts. Addictions aren't driven by substances or behaviors, but by the idea, learned across time, that they protect addicts from psychological distress.

The truth about addiction challenges many of our intuitions. It isn't the body falling in unrequited love with a dangerous drug, but rather the mind learning to associate any substance or behavior with relief from psychological pain. In fact, addiction isn't about falling in love; as Kent Berridge showed, all addicts *want* the object of their addiction, but many of them don't *like* it at all. As for Isaac Vaisberg, Andrew Lawrence's Parkinson's patients, and Rat No. 34, addiction persists even after its appeal wanes, leaving intact the desire for gaming, tidying up obsessively, or self-administering a shock long after the pleasure has gone.

PART 2

The Ingredients of Behavioral
Addiction (or, How to Engineer
an Addictive Experience)

4.

Goals

In 1987, three Australian neurologists stumbled upon a simple technique that improved the lives of thousands of Parkinson's patients. The disease renders many patients unable to walk as tremors cause them to freeze in place. The neurologists began their report by describing an accidental discovery. A man who had suffered from Parkinson's for eleven years was still able to stand from a seated position, but he could no longer walk. One morning he swung his legs over the side of his bed, and planted them firmly on the ground. He rose, and once he was standing looked down and noticed that his shoes sat neatly just in front of his feet, like two small hurdles. Much to his surprise, instead of struggling to walk, he was able to take a tentative step over the one shoe, and then a second step over the other. The shoes were now behind him. With the help of the bite-sized goal of stepping over his shoes, he had managed for the first time in years to walk rather than shuffle.

The man was enterprising. He experimented with different techniques. First he carried small objects wherever he went, tossing them a couple of inches ahead whenever he froze in place. Soon you could trace his path around the house by following the trail of household items. Then, overwhelmed by the motley detritus strewn across his floor, he discovered that he could use his walking stick as a reusable hurdle. He turned the walking stick upside down so its handle met the ground just in front of his right foot. With the handle as a hurdle, he took a first step, and then repeated the tactic with his left foot. Having gained some momentum after a couple of steps, he established a regular gait and managed to walk slowly without the aid of his stick.

The man visited his neurologist, one of the three who went on to write the landmark paper, and demonstrated his new trick. The neurologist was floored. How was it possible that a hurdle had improved his patient's gait? The answer is that, if you want to compel people to act, you whittle down overwhelming goals into smaller goals that are concrete and easier to manage. Humans are driven by a sense of progress, and progress is easier to perceive when the finish line is in sight. Using his walking stick, the man created a series of bite-sized goals that were easy to digest. When the neurologist and two of his colleagues confirmed that the approach worked for other Parkinson's patients, they described it in the paper that armed neurologists with a new tool to treat one of Parkinson's most debilitating symptoms.

Like small hurdles to a Parkinson's patient, goals often inspire action because they become fixation points. You can see this when you examine the finishing times of millions of goal-driven marathon runners.

———

The average marathon runner completes the 26.2 mile (42.2 kilometer) course in roughly four and a half hours. At the extremes, elite male athletes run the distance in just over two hours, while the slowest walkers spend ten or more hours on the course. You might expect a smooth distribution of times between those extremes. Something like this, where the height of each bar indicates how many runners finished with that time:

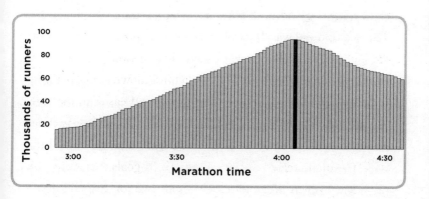

A small group of runners finish in less than three hours, with slower times becoming more common until a peak at four hours and three minutes (the dark bar). There aren't any obvious jumps or troughs in the distribution, which is typical of how humans perform many physical tasks.

But that's not how the distribution actually looks, because certain milestone times are more meaningful than others. I know this from experience, because I ran the New York City Marathon in 2010. Many athletes run behind pacesetters who carry large placards displaying times like "3:00," "3:30," or "4:00." The pacesetters who carry those signs are experienced runners who aim to finish just under those milestones, and they usually succeed. I followed the 3:30 pacesetter for as long as I could, but as the race wore on I slowed down. When the 3:30 pacesetter was so far ahead that I could barely read his sign, the 4:00 pacesetter came up alongside me. I abandoned my race plan and set a new, firm goal: this might be my only marathon, so I absolutely had to finish in under four hours. Now a few miles from the end of the race, I was utterly depleted. I remember wolfing down a couple of bananas offered by a kind spectator who took pity on me. A friend jumped onto the course and shouted, "Good job! You're right on pace to finish in under four hours and five minutes!" His words exposed a hidden well of energy, and I ran slightly faster for the remainder of the race. My time: 3:57:55. When I found the same friend after the race he told me he had lied. "You were on track to finish in just under four hours, but I was worried you might slow down," he told me. "I knew you'd dig deeper if you thought you were on track to finish in 4:05." The 2010 New York City Marathon was my first and last so far, but I'd have run the race again in 2011 had my finishing time exceeded four hours.

I'm not alone. In 2014, four behavioral scientists plotted the finishing times of almost ten million marathon runners on a single graph:

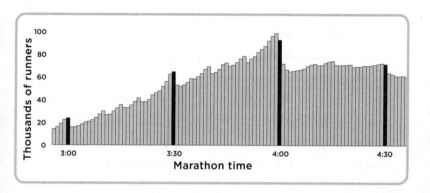

If you focus on the milestone times that arrive every half hour, you can see my struggle writ large. The dark bars indicate times just below those milestones (2:59, 3:29, 3:59, and 4:29), and you can see they're a lot more common than slightly slower times (the two or three shorter bars to their right). Runners somehow find buried stores of energy as they strive to beat significant milestones, so many more of them finish in, say, 3:58 or 3:59 than in 4:01 or 4:02. In a race like the New York City Marathon, with almost fifty thousand runners, five hundred will finish with a time of 3:59, while just three hundred and ninety will finish with a time of 4:01. The size of that difference captures how urgently marathon runners want to finish in less than four hours. That's the compelling power of goals: even when you're two bananas shy of collapsing, you find the will to go on. So what happens when you reach your goal?

Robert Beamon was born into a poor family in Queens, New York, just after the end of World War II. His father was abusive and his mother feared for Bob's life, so young Bob was sent to live with his grandmother. By the time he attended high school, he was tall, wiry, and athletic. He liked running and jumping, and a scout took notice as he bounded down the long jump runway. He began to win U.S. high school titles, and by the time he graduated from high school he was among the best two or three long jumpers in the country. Beamon accepted a scholarship to the University of Texas at El Paso, where he adopted the ultimate goal for elite athletes: to win Olympic gold.

Beamon's opportunity came in 1968, at the Mexico City Olympic Games. He arrived in Mexico City having won twenty-two of his previous twenty-three events, and was hotly favored to win gold. But Beamon panicked. During the qualifying rounds, something didn't feel right. For the first time in years he was wracked with nerves. Out of sorts, Beamon misjudged his run-up, and fouled his first two jumps. Everything rode on his third and final chance to qualify. Ralph Boston, the reigning world record holder and Beamon's teammate, pulled Beamon aside and told him to jump conservatively. In Beamon's memory, "he said, 'pull it back three feet, and if you have to jump behind the board, jump behind the board.'" On his third and final jump, despite leaping from far behind the board, Beamon managed to qualify.

The final began the next morning. When Beamon reflected on the event in an interview, forty years later, he remembered feeling "calm, very peaceful." He later told interviewers he had thrown back a couple of tequila shots the night before, briefly abandoning his abstemious training regime. Three athletes were listed to jump before Beamon, but all three fouled their initial jumps, so Beamon was left without a target distance. His first jump took just seven seconds from beginning to end. He tore down the runway and leapt high and far, coming down a very long way from the beginning of the sandpit. Beamon leapt so far that the electronic measurement system was incapable of calculating the jump's distance. You can still watch the footage today: an earnest official moves the measuring device to the end of a fixed railing, and allows himself a brief smile when he realizes the jump is unmeasurable. The officials confer briefly before they realize there isn't a suitable measuring tape in the stadium. One of them is dispatched to find a tape while the event is suspended. Forty-five minutes pass, a tape is found, and after measuring and remeasuring, the officials put a number to Beamon's colossal leap: eight meters and ninety centimeters, or twenty-nine feet, two and a half inches. Beamon had jumped almost two feet—fifty-five centimeters—farther than any other man in history. He collapsed to the track as Boston tried to help him to his feet, only to collapse again before his legs could support his weight. Watching the footage, one doctor diagnosed Beamon with a cataplectic seizure brought on by the emotional shock of his achievement. The jump was so impressive that the word *Beamonesque* came to signify an otherworldly achievement that dwarfs its predecessors.

Beamon had demolished his athletic goals. He was an Olympic gold medalist and a world record holder. A dejected Lynn Davies, the defending Olympic long jump champion, asked,

"What's the point? Beamon's jump has destroyed the event." To Russian champion Igor Ter-Ovanesyan, Beamon's competitors were mere "children." The record stood for almost twenty-three years, until U.S. athlete Mike Powell surpassed Beamon's mark by a couple of inches—a record that still stands today.

In theory Beamon should have been elated. After a shaky qualifying round, he managed to summon a performance regularly cited among the five greatest athletic feats of all time. The remainder of that day should have been one of the greatest of Beamon's life. But that's not what happened. In 2008, he remembered his celebration lasting only a few minutes. "When I got to the medal stand, I said, 'what am I gonna do? I've reached one stage, and so what is the next peak experience in my life?'"

A week after the event he was taking sociology classes toward a master's degree at Adelphi University. He all but abandoned athletics, and even today you can see when he's asked about the feat that it doesn't bring him much joy. He summons a brief nod, quietly admits the jump was impressive despite his modest disposition, and then goes on to discuss his work as a philanthropist or the virtues of his Olympic teammates.

Beamon is perhaps unusually reserved, but even flamboyant goal-setters struggle with outrageous success. That was certainly the case for Michael Larson, a legend in the world of TV game shows.

———

Larson was known as a goal guy. He pursued goals constantly, sometimes small ones and sometimes big ones. Many of them involved making money, because Michael was born in 1949 to a family of modest means in small-town Lebanon, Ohio. Sometimes his behavior wasn't strictly ethical. That was true of the time he quietly sold candy bars at a profit to his middle school classmates, and of the time he opened up multiple bank accounts under different names so he could claim the bank's five-hundred-dollar new account bonus. Larson was a serial goal monogamist, lining up one goal after another and rarely attacking a new one until his current project was complete. He was rarely without a quest, and he was restless without a new one on the horizon. It was this fevered approach to goals that triggered his undoing.

In the summer of 1983, Larson was thirty-four years old and unemployed, apart from scattered work as an ice cream truck driver and air conditioner repairman. He assembled a huge wall of TVs and obsessively scoured the networks for moneymaking opportunities. Eventually, he found his mark in the form of a new game show. *Press Your Luck* premiered on CBS in September 1983. The show's premise was simple: contestants answered trivia questions to accumulate "spins," and then used those spins on a big game board to win cash and prizes while avoiding "whammy" squares that returned their winnings to zero. Contestants watched as a flashing light darted around the game board's eighteen squares, and pressed a red button to stop the light wherever it happened to be at the time. The content of each square changed from moment to moment, making it very difficult to predict whether the selected square would depict cash, prizes, or a whammy. Hence the show's name: contestants could choose to pass their remaining spins to the next player, or continue to "press their luck" by taking another turn. The game was designed so the light landed on a whammy every five or six spins, preventing players from stringing together more than a handful of winning spins in a row.

Larson watched the game with great interest. To most people the board seemed to behave randomly, but Larson wasn't like most people. He sat, day after day, recording the outcome of each spin until a series of patterns resolved before him. He shared the discovery with his wife: the flashing light followed five distinct patterns, and no matter which pattern it followed, two of the eighteen squares never displayed a whammy. In one pattern, for example, the flashing light landed on a safe square after landing on four dangerous squares. With practice, anyone could learn to game the system. Larson had a new goal.

For six months, Larson memorized the five winning patterns, playing along with the contestants until he was eating, sleeping, and breathing the magic sequences. He numbered each square and rehearsed the light's path as it bounced around the board. "Two. Twelve. One. Nine. Safe! Two. Twelve. One. Nine. Safe!" His behavior was eccentric, sure, but Larson was willing to go to great lengths to achieve his potentially lucrative goal.

One day, Larson told his wife he was ready. He gathered every penny he had and traveled from Ohio to the *Press Your Luck* studios in Los Angeles. He wore a rumpled gray suit on the plane, and then wore that same suit every morning and afternoon as he auditioned twice a day for several days along with fifty other hopefuls. His optimistic energy charmed enough of the casting crew that they invited him to appear on the show on May 19, 1984.

The show began as it did most days. Peter Tomarken, its affable host, asked Michael what he did for a living, and joked that, though Michael had probably overdosed on ice cream as an ice cream truck driver, he hoped Michael wouldn't overdose on money. When the trivia round began, it became clear that Larson was different from his two competitors. While they pushed their red buzzers casually with one hand, Larson used a two-handed grip and struck the buzzer like a rattlesnake. Here was a man with technique, a man who had spent months planning his conquest.

But Larson's bid didn't begin as planned. His first spin turned up a whammy. Apparently the game board took a fraction of a second to react to the buzzer. Larson was briefly dazed, but soon hit his stride and began to amass a mountain of cash and prizes. The show's associate director, Rick Stern, recognized the determined look on Larson's face. "I have a fifteen-year-old son who plays video games, and that's the look on his face when he enters the zone. Larson was looking for his patterns, and he had a lot of work to do." Adrienne Pettijohn, a production assistant, only half joked that "this guy's going to walk away with the network."

With each successful spin, Larson yelped with glee. Four thousand dollars and a free spin. Five thousand dollars and a free spin. A vacation in Kauai. One thousand dollars and a free spin. A sailboat. And so on. On Larson's left, contestant Ed Long began to cheer as he too was swept away by Larson's improbable run. On Larson's right, Janie Litras grew angrier with each spin. Reflecting on her loss two decades later, she remembered, "I wasn't into it. I was getting madder and madder. I was supposed to be the winner."

Larson ignored Long and Litras as his winnings soared. Ten thousand dollars. Twenty thousand dollars. At twenty-six thousand, Tomarken shouted, "Unbelievable! What's happening here?" Behind the scenes, the show's executive staff began to panic. A year earlier, while designing the game, they had dismissed the possibility that some enterprising contestant might learn the board's five pre-programmed patterns. Meanwhile, instead of passing his remaining spins, Larson pressed on—past thirty thousand dollars, and then forty thousand dollars, and then past forty-four thousand dollars, the highest single-day winnings the show had seen to date. Then on past fifty, sixty, and seventy thousand dollars—and past the highest single-day winning total on *any* American game show with returning champions.

By all rational accounts, Larson should have stopped. One whammy would have ended his run and left him with precisely nothing—a colossal loss of tens of thousands of dollars. Ignoring Tomarken's gentle warnings, Larson became obsessed with a magic total. "I'm going for a hundred!" he shouted shortly after his thirtieth winning spin. When he reached a hundred thousand dollars, the score board jettisoned its dollar sign; it had been designed to max out at $99,999.

And then the wheels almost came off. Larson was two spins from victory when his concentration slipped. Instead of hitting one of the safe squares, the light stopped on a dangerous square. Larson had allowed the light to move one square too far. On his first spin this square had displayed a whammy, but this time the gods were smiling: seven hundred and fifty dollars and a free spin. Larson was shaken, but he pressed on with his final spin and turned up a trip to the Bahamas. The result: a total of $110,237 in winnings, to this day more than any other contestant has won on a single episode of any game show with returning champions.

After Larson's performance, the *Press Your Luck* executives revised the game's mechanics so the board alternated among thirty-two different sequences instead of just the original five. At the same time, they eliminated safe squares—depending on the sequence, any square could contain a whammy. Now it was almost impossible for a contestant to predict where the light might jump next, and what it might illuminate.

And what of the victorious Michael Larson? The CBS team tried to argue that Larson had cheated, but in truth he had done nothing wrong at all. Reluctantly they paid the amount in full, and Larson returned to Ohio a wealthy man. By all accounts, he had exceeded every possible goal he might have had when he boarded the plane to Los Angeles: no other game show contestant had ever won as much on a single day, and he had won more than a hundred thousand dollars. But just as Larson had refused to pass his spins on the show, so he refused to rest on his laurels at home.

Still restless, Larson grew addicted to a goal that destroyed his marriage and left him penniless. A local radio station offered to pay a lucky listener thirty thousand dollars for sending in a dollar bill with a serial number that matched the number randomly called out on the air each day. Serial numbers are eight digits long, so the chances of winning this particular lottery are vanishingly small—roughly one in a hundred million. Larson mistakenly believed it was only a matter of time till he won if he converted the remaining fifty thousand dollars of his *Press Your Luck* winnings into one dollar bills—a total of fifty thousand chances to win. Each day, as the radio show called out the winning serial number, Larson and Teresa sat for hours and leafed through the pile of bills. Teresa grew to despise him. He was so focused on the game that he became distant and bitter.

One night the couple went to a Christmas party, and a band of thieves broke in, stealing all but five thousand dollars of Larson's winnings. Teresa was so angry that she absconded with the five thousand dollars, and never saw Michael again. Soon afterward, he moved to Florida, living the remaining fifteen years of his life in pursuit of increasingly dubious schemes. Larson became a tragic emblem for goal addicts everywhere: mountaineers who refuse to stop climbing new peaks even in the face of death, gamblers who refuse to stop betting as their lives crumble, and workers who refuse to go home even if they have no need to work more.

Bob Beamon and Michael Larson differ in so many ways. Beamon overachieved and Larson was a serial underachiever. Beamon is modest and reserved, Larson was flamboyant and naïvely candid. But both of them sacrificed immediate well-being for the promise of long-term success, and were surprised when their immense achievements brought them very little joy. Like the curse that doomed Sisyphus to roll a boulder uphill for eternity, it's hard not to wonder whether major life goals are by their nature a major source of frustration. Either you endure the anticlimax of succeeding, or you endure the disappointment of failing. All of this matters now more than ever because there's good reason to believe we're living through an unprecedented age of goal culture—a period underscored by addictive perfectionism, self-assessment, more time at work, and less time at play.

Despite all the drawbacks of goal-setting, the practice has increased in the past several decades. What is it about the world today that makes goal pursuit so alluring?

Goals have been around for as long as our planet has sustained life. What has changed, though, is how much of our lives are occupied by goal pursuit. Once upon a time goals were mostly about survival. We foraged for food and preened for attractive mates, and these activities were critical to the survival of our species. Goals were a biological imperative rather than a luxury or a choice. Our species would never have survived had our ancestors spent their time pursuing goals for no good reason. When food and energy were scarce, the guy who climbed the nearest mountain just for fun, or ran a hundred miles just to see if he could, didn't last very long at all. Today, for much of the world, food and energy are abundant, and you can live a long and happy life while choosing to take on unnecessary hardships like mountaineering and ultramarathon running. And once you've finished climbing one mountain or running one race, you can start preparing for the next one, because today goals are far more than just destinations; today we're fixated on the journey, and often the act of reaching the goal is an incidental anticlimax.

There's plenty of evidence for this rise in goal culture if you know where to look. You can see it in the rise of the phrase "goal pursuit," which was absent from English language books until 1950:

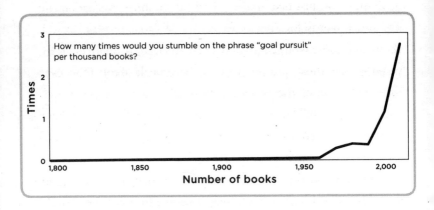

The concept of setting one goal after another—of perfectionism—is also quite new. The word barely existed in the early 1800s, but it seems to be everywhere now. In 1900 the word appeared in just 0.1 percent of every book (you'd need to read more than one thousand books to see it written just once). Today roughly 5 percent of all books (or one in twenty) mention the idea of "perfectionism."

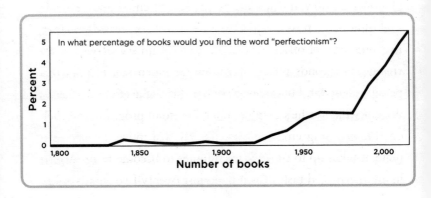

This could just be a matter of language shifts; maybe people had other words for "perfectionism" and "goal pursuit" in the 1800s, and those words have now been replaced. If that were true, you'd expect those phrases to have become less common over time, but none of the dictionary synonyms for "perfectionism" and "goal pursuit" have died out. If anything, most of them have become more common—terms like "quest," "plan," target," "objective," and "striving."

Even beyond the world of books, goals have become harder to escape. The Internet has exposed people to goals they barely knew existed, and wearable tech devices have made goal tracking effortless and automatic. Where once you had to seek out new goals, today they land, often uninvited, in your inbox and on your screen. We might get by if we were able to leave those emails unread for hours or even days at a time, but to the detriment of productivity and well-being, we can't help responding to new emails almost as soon as they arrive.

How long do you think the average office email goes unread? I guessed ten minutes. The truth is just six seconds. In reality, 70 percent of office emails are read within six seconds of arriving. Six seconds is less time than it's taken you to read this paragraph so far, but it's long enough for the average worker to disrupt whatever he's doing to open his email program and click on the incoming email. This is hugely disruptive: by one estimate, it takes up to twenty-five minutes to become re-immersed in an interrupted task. If you open just twenty-five emails a day, evenly spaced across the day, you'll spend literally no time in the zone of maximum productivity.

The solution is to disable new email notifications and to check your email account infrequently, but most people don't treat email that way. Many of us pursue the unforgiving goal of *Inbox Zero*, which requires you to process and file away every single unread email as soon as it arrives. And, as Chuck Klosterman wrote in the *New York Times*, emails are like zombies: you keep killing them and they keep coming. Inbox Zero also explains why workers spend a quarter of their days dealing with emails, and why they check their accounts, on average, thirty-six times every hour. In one study, researchers found that 45 percent of respondents associated email with "a loss of control." This from a mode of communication that barely existed until the twenty-first century.

In 2012, three researchers wanted to investigate what happens when you prevent office workers from using email for a few days, but they struggled to find volunteers. They approached dozens of office workers at a U.S. Army facility on the East Coast, but only thirteen were willing to participate in the study. The vast majority explained that they couldn't bear the pain of sorting through hundreds of unanswered emails when the study ended. Inbox Zero never dies; it just grows angrier while you try to ignore it.

The researchers monitored the thirteen volunteers for eight days in total: three days as they continued using email as they usually did, and then five days while they refrained from using email altogether. At first the volunteers felt disconnected from their workmates, but quickly took to walking around the office and using their desk phones. They also left the office more often, spending three times as long outside when they were forbidden from using email. Apparently email kept them shackled to their desks. They were also better workers, switching between tasks half as often, and spending longer on each task without distraction. Most important, though, they were healthier. When checking email, they were in a constant state of high alert; without email, their heart rates tended to vary more, rising in response to brief bursts of stress, but falling again when those stressors passed. With email they were constantly on red alert.

Beyond Inbox Zero, the Internet has also made it easier to stumble on new goals. Even just twenty-five years ago, goals were more remote than they are today. My family moved from Johannesburg, South Africa, to Sydney, Australia, when I was seven. Two months later my grandma visited from South Africa to help us settle in. As always, she brought gifts, and one of those gifts was the 1988 edition of the *Guinness Book of World Records*. After I tore open the wrapping, she directed me to a section titled "Human Superlatives." There, on the left-hand page, was a picture of Robert Pershing Wadlow, the tallest man of all time. At his peak, Wadlow stood 8 feet, 11.1 inches tall. "I met that man when he visited South Africa," my grandma told me. "I was a child, but I remember standing next to him as he looked down and smiled." I was hooked. I read the book over and over. I memorized the size of Wadlow's shoes (37AA), the weight of the world's heaviest man (1400 pounds), and the greatest number of lightning strikes survived by one person (seven, by a park ranger named Roy Sullivan). The records were exotic and remote, which is precisely why I found them so fascinating.

Today, records and goals are everywhere, and anyone can

participate in the act of record-setting—a symptom of the age of information. The Guinness World Records homepage features a button labeled SET A RECORD. Follow the link and you'll see the smiling faces and medaled chests of recent record-setters. Gunnar Garfors and Adrian Butterworth visited five continents in one calendar day. Hiroyuki Yoshida and Sandra Smith got married 130 meters underwater. Steve Chalke has raised millions of pounds for charity while running marathons, more than anyone else in history. And so on. It's never been so easy to concoct a goal—and, much to our detriment, we're coaxed along that complicated path by devices that are meant to make our lives easier.

Katherine Schreiber and Leslie Sim are experts on exercise addiction who believe that tech advances encourage obsessive goal monitoring. Schreiber and Sim loathe wearable tech. "It's the worst," Schreiber says. "The dumbest thing in the world," says Sim. Schreiber has written extensively about exercise addiction, and Sim is a clinical child adolescent psychologist at the Mayo Clinic. Many of Sim's adolescent patients have twin exercise and eating disorders, which tend to go together.

Wearable tech is a catchall term that describes clothing and accessories with electronic computer-based functions. Websites like Guinness World Records made goals more prominent, but they have nothing on wearable tech. Schreiber and Sim were particularly critical of watches and trackers that present wearers with instantly updated fitness metrics. Many of these devices either give you goals or ask you to nominate your own. The gold standard is step milestones, or the number of steps the wearer walks each day. Reach the goal—ten thousand steps, for example—and the device emits a reinforcing beep. I've watched friends and family members respond to that beep, and it's hard not to think of Pavlov's dog.

Schreiber and Sim both recognized that smartwatches and fitness trackers have probably inspired sedentary people to take up exercise, and encouraged people who aren't very active to exercise more consistently. But as experts in addiction, they were convinced the devices were also quite dangerous. Schreiber explained that "focusing on numbers divorces you from being in tune with your body. Exercising becomes mindless, which is 'the goal' of addiction." This "goal" that she mentioned in quotes is a sort of automatic mindlessness, the outsourcing of decision making to a device. She had recently sustained a stress fracture in her foot because she refused to listen to her overworked body, instead continuing to run toward an arbitrary workout target. Schreiber has suffered from addictive exercise tendencies, and vows not to use wearable tech when she works out.

I use a watch that tracks my progress when I run outdoors, and I hate to stop until I hit a predetermined number of whole miles. Occasionally the watch won't work, and those runs, untethered to numbers, are always my favorite. In a *New Yorker* piece, the humorist David Sedaris described how owning a Fitbit changed his life.

> During the first few weeks that I had it, I'd return to my hotel at the end of the day, and when I discovered that I'd taken a total of, say, twelve thousand steps, I'd go out for another three thousand.

"But why?" [my husband] Hugh asked when I told him about it. "Why isn't twelve thousand enough?"

"Because," I told him, "my Fitbit thinks I can do better."

I look back at that time and laugh—fifteen thousand steps—Ha! That's only about seven miles! Not bad if you're on a business trip or you're just getting used to a new prosthetic leg.

Numbers pave the road to obsession. "When it comes to exercise, everything can be measured," Sim says. "How many calories you burn; how many laps you run; how fast you go; how many reps you do; how many paces you take. And if you went, say, two miles yesterday, you don't want to go less than that today. It becomes fairly compulsive." Many of Sim's patients experience this constant need to check in. A ten-year-old boy who visited her clinic in Minneapolis was known for being a fast runner, and he wore his speed as a badge of honor. His biggest concern was that he might slow down, so he constantly checked by moving all the time. "He would drive his parents crazy. When they visited Minneapolis for their evaluation, he kept the entire hotel awake at night. They were getting complaints because he was running around his room."

Sim's patient was obviously in psychological distress, but most people become obsessive when they're focusing on numbers. "Counting steps and calories doesn't actually help us lose weight; it just makes us more compulsive. We become less intuitive about our physical activity and eating." Even if you're tired, and feel you need to rest, you'll continue walking or running till you reach your arbitrary numerical goal. Schreiber agreed. To her, the pangs she feels when she's not working out are a lot like love. "When you're not with your significant other, or whoever you're in love with, you long to be with that person." The moral is that it's healthy to make goals more difficult to measure, but also that it is dangerous to have devices that monitor everything from our heart rates to the number of steps we've walked today.

Schreiber's love affair with running isn't unusual. In 2000, Marylanders Dawn and John Strumsky founded the United States Running Streak Association (USRSA). The association celebrates runners who haven't missed a day of running for many years. ("Running" consists of traveling one mile or more without the aid of crutches or a stick.) It's a tremendously supportive group, typical of community-minded running organizations. The USRSA hosts a diverse mix of runners—young and old, male and female, elite and non-competitive—who are united by their drive to run every single day. The association releases a quarterly bulletin that celebrates milestones. Run for thirty-five years straight and you become a Grand Master; forty years and you become a Legend. If you reach forty-five years, you're called a Covert, after Mark Covert, who retired when he became the first person to reach the forty-five-year mark, in 2013.

As you can imagine, many of the association's runners have persevered through near-impossible conditions. When Gaby Cohen discovered she needed a C-section some years ago, she found a private hospital bathroom and ran in place for twelve minutes. Cohen passed the twenty-two-year mark in November, 2014. (Cohen's streak is impressive, but a sixty-three-year-old Californian named Jon Sutherland holds the U.S. record at forty-six years, and counting.) When Hurricane Frances passed directly over David Walberg's hometown in 2004, he waited till its calm eye winds arrived before squeezing in a 1.2 mile run. Walberg's streak stands at thirty-one years. Other runners blaze down airport corridors when their flights are canceled, and persevere through debilitating illness and injury. Anything to keep the streak alive.

There's also something insidious about streaks. Because they demand repeated activity without a break, they become more precious over time. A two-week streak isn't much to protect, but even laid-back runners slavishly protect streaks that reach beyond the one-year mark, running on a hobbled ankle or through a bout of the flu. Robert Kraft, a sixty-four-year-old runner from Miami, recently hit the forty-year mark. Kraft pushes through arthritis, a painful condition that affects his spine, and a degenerated disk. Each run is painful for Kraft, but he wouldn't dream of missing a day. This is dangerous and even the Running Streak Association website now publishes a warning, written by founder John Strumsky, imploring streakers to "rest and recuperate to avoid injuries." To most runners, this means a day of rest, but to streakers, it's a day of easy running. For many people, the heaviest cost of sustaining a streak is psychological. After compiling a streak of one hundred and thirty-one days, Michelle Fritz realized the streak was "becoming an idol." She had no time for her husband and children, and decided to skip a day. "I really felt better after ending it," she recalled, though she's now one hundred days deep into a new streak. Old goals, it turns out, die hard.

Streaks uncover the major flaw with goal pursuit: you spend far more time pursuing the goal than you do enjoying the fruits of your success. Even if you succeed, success is brief. Writing for the *Guardian*, human behavior expert Oliver Burkeman explained:

> When you approach life as a sequence of milestones to be achieved, you exist "in a state of near-continuous failure." Almost all the time, by definition, you're not at the place you've defined as embodying accomplishment or success. And should you get there, you'll find you've lost the very thing that gave you a sense of purpose—so you'll formulate a new goal and start again.

Burkeman was quoting from Scott Adams, the cartoonist and creator of the *Dilbert* comic strip, who condemned goal pursuit in his book, *How to Fail at Almost Everything and Still Win Big*. Adams promoted an alternative: instead of goals, live your life by systems. A system is "something you do on a regular basis that increases your odds of happiness in the long run." For a cartoonist, that might be drawing one cartoon per day; for a writer, writing five hundred words per day. In contrast to goals, systems bring a steadier stream of low-grade highs. They're guides to a fulfilling life, day by day, rather than enticing pictures of some grand end goal without instructions for how to get there.

Systems stand in stark contrast to goals like "attract one thousand Instagram followers," which serve only as signposts of failure. When you do reach your goal, a new one materializes in its place—now two thousand Instagram followers seems like an appropriate target. The defining goal of our time, perhaps, is to amass a certain sum of money. That sum begins small but grows over time. In 2014, a former Wall Street trader named Sam Polk published an op-ed in the *New York Times* titled "For the Love of Money." Polk explained that his goal was modest, at first, and then escalated repeatedly. "I'd gone from thrilled at my first bonus—$40,000—to being disappointed when, my second year at the hedge fund, I was paid 'only' $1.5 million." Some of Polk's bosses were billionaires, so he, too, wanted a billion dollars. "On a trading desk, everyone sits together, from interns to managing directors," Polk said. "When the guy next to you makes $10 million, $1 million or $2 million doesn't look so sweet."

Polk was describing the principle of social comparison. We constantly compare what we have to what other people have, and the conclusions we draw depend on who those people are. A bonus of $40,000 looks terrific when you remember that some of your friends earn $40,000 a year; but if your friends are high-flying traders who earn $40,000 a week, you'll be disappointed. Humans are inherently aspirational; we look ahead rather than backward, so no matter where we stand, we'll tend to focus on people who have more. That experience produces a feeling of loss, or deprivation, relative to those other people. That's why Polk was never happy; no matter how much he earned, there was always someone who earned more. As ridiculous as it may sound, even billionaires are poor next to multibillionaires, so they, too, feel the sting of relative deprivation.

I asked Polk whether his experience was common. "I think it's 90-plus percent pervasive in finance, and I also think it goes way beyond finance." Polk reminded me of a recent Powerball draw that attracted millions of entrants vying for a colossal $1.6 billion jackpot. Polk was convinced that this perpetual goal, even among the very wealthy, reflected a "lack of connection with your life's work." You don't need to keep score with money if you're truly, deeply motivated by what you're doing. Goals function as placeholders that propel you forward when the daily systems that run your life are no longer fulfilling. Echoing Adams and Burkeman, Polk told me that the key is to find something that brings you small doses of positive feedback. He also believes that wealth addiction is a relatively new phenomenon. In his 1989 book, *Liar's Poker*, Michael Lewis, a former trader himself, wrote that traders once believed they were performing a social function. They funded important projects and made sure money traveled from where it was to where it could be more useful. It fueled the construction of buildings and industry, and created thousands of jobs. But that illusion has gone, Polk says, as has the intrinsic motivation to trade for anything but personal gain. In 2010, Polk left Wall Street behind, choosing instead to write a book, and to found a food nonprofit called Groceryships.

In moderation, personal goal-setting makes intuitive sense, because it tells you how to spend your limited time and energy. But today, goals visit themselves upon us, uninvited. Sign up for a social media account, and soon you'll seek followers and likes. Create an email account, and you'll forever chase an empty inbox.

Wear a fitness watch, and you'll need to walk a certain number of steps each day. Play Candy Crush and you'll need to break your existing high score. If your pursuit happens to be governed by time or numbers—running a marathon, say, or measuring your salary—goals will come in the form of round numbers and social comparisons. You may find you want to run faster and earn more than other people, and to beat certain natural milestones. Running a marathon in 4:01 will seem like a failure, as will earning $99,500. These goals pile up, and they fuel addictive pursuits that bring failure or, perhaps worse, repeated success that spawns one new ambitious goal after another.

5.

Feedback

Last week I stepped into an elevator on the eighteenth floor of a tall building in New York City. A young woman inside the elevator was looking down at the top of her toddler's head with embarrassment as he looked at me and grinned. When I turned to push the lobby button, I saw that every button had already been pushed. Kids love pushing buttons, but they only push *every* button when the buttons light up. From a young age, humans are driven to learn, and learning involves getting as much feedback as possible from the immediate environment. The toddler who shared my elevator was grinning because feedback—in the form of lights or sounds or any change in the state of the world—is pleasurable.

This quest for feedback doesn't end at adulthood. In 2012, an ad agency in Belgium produced an outdoor campaign that quickly went viral. The agency, Duval Guillaume Modem, was trying to convince the Belgian public that the TNT television network was broadcasting shows that were packed with excitement. The campaign's producers placed a big red button on a pedestal in a quaint square in a sleepy town in Flanders. A big arrow hung above the button with a simple instruction: *Push to add drama.* The campaign worked beautifully because buttons, even in quiet Flemish squares, beg to be pushed. (The sign was a nice but unnecessary touch—with mounting curiosity, people will eventually push a big conspicuous button even if it isn't labeled.) A couple of adults sidled up to the button before braver souls went one step further and pressed down. You can see the glint in each person's eye as he or she approaches the button—the same glint that came just before the toddler in my elevator raked his tiny hand across the panel of buttons. (The ad's YouTube video has more than fifty million hits. As promised by the arrow, the result is dramatic, featuring bumbling paramedics, a fistfight, a bikini-clad woman on a motorbike, and a shootout.)

The button in Flanders promised a reward, but people will also push buttons that promise nothing at all. This was the case when the Reddit web community posted an April Fools' Day prank in 2015. Reddit, which celebrated its tenth birthday in June 2015, is currently the thirtieth most popular website on the Internet, attracting slightly more traffic than Pinterest and slightly less traffic than Instagram. It houses a motley collection of pages devoted to news, entertainment, and social networking. Users celebrate some posts by clicking upward-pointing arrows, and condemn others by clicking downward-pointing arrows. Each post features a running score that rises and falls as users bestow these upvotes and downvotes. To give you a sense of Reddit's irreverence, one of the most upvoted posts of all time is titled "Waterboarding in Guantanamo Bay sounds rad if you don't know what either of those things mean."

On April 1, 2015, Reddit unleashed a prank on its thirty-five million registered users. One of the site's administrators introduced the prank in an announcement on the Reddit blog page:

the button
Posted by the reddit admins at 09:00 | 🔆 ⬆ ⬇ 0 points | ✉
Labels: **activate**, **bear down on**, **depress**, **operate**, **press**, **squeeze**

dramatization

reddit,

When this post is 10 minutes old, a button and timer will become active at /r/thebutton. The timer will count down from 60 seconds. If the button is pressed the timer will reset to 60 seconds and continue counting down. Only users logged into accounts created before 2015-04-01 can press the button.

You may only press the button once.

We can't tell you what to do from here on out. The choice is yours.

visit /r/thebutton

The button's mechanics were simple: a timer next to the button would descend from sixty seconds to zero. Every time a user clicked the button, the counter would return to sixty seconds to restart its downward march. Users could only click the button once, so the timer would reach zero eventually. (Even if every one of Reddit's subscribers clicked the button just before it reached zero, the timer would reach zero after sixty-six years.)

At first, hordes of users visited the page and, almost without fail, pushed the button before it had descended much below sixty seconds. These users received a small purple badge next to their usernames, with a number that indicated how many seconds were left on the countdown timer when they clicked. Users who were especially trigger-happy had purple badges that broadcast "59 seconds"—a number that suggested the user was impatient. The button didn't appear to do much, apart from turning the badge purple, so it wasn't clear why some users were staying up all night to wait for the timer to fall. Such was the lure of the button—like elevator buttons to toddlers—that they were willing to forgo sleep for the chance to push the button at a low number.

Interest in the campaign surged. Users who hadn't yet pressed the button had gray badges, and many of them counseled other gray-badged users to join the "don't press!" camp. If enough people refused to press the button, they reasoned, it would reach zero more quickly, and the result of the campaign would reveal itself sooner. But hundreds of thousands of users couldn't resist the urge to press, and the timer crept down very slowly. On April 2, the timer reached fifty seconds for the first time, and the user who pressed the button received a blue badge. All users who clicked the button when it had fallen below fifty-one seconds received a blue badge instead of a purple one. Users quickly learned that their reward for clicking the badge as it dropped below each ten-second interval was a different-colored badge—not a huge reward, perhaps, but users formed camps based on the color of their badges, and later pushers wore their badges with special honor. Here's a full list of how long it took for the clock to reach each badge, and how many users earned each one:

Badge color	When badge was activated	Percent of clickers who earned this badge	When badge was first earned
Purple	52–60 seconds	58	April 1
Blue	42–51 seconds	18	April 2
Green	32–41 seconds	8	April 4
Yellow	22–31 seconds	6	April 10
Orange	22–21 seconds	4	April 18
Red	Below 11 seconds	6	April 24
Purple (again)	Final presser	One user: BigGoron	May 18

As the colors revealed themselves, one Reddit user named Goombac created avatars for each camp, and christened them with names like *The Illemonati* (yellow, of course), *The Emerald Council*, and *The Redguard*. Forty-eight days after the prank began, BigGoron pressed the button for the last time. After his push, the countdown timer descended to zero. Reddit hailed Big-Goron the *Pressiah*, and users bombarded him with questions. How had he waited when so many before him had fallen? (He noticed that the timer had reached one second a number of times, so he began to watch and wait.) What comes next? ("I advocate peace—please let the crusades end.") In the end, when the timer reached zero, nothing happened at all. Users formed camps united by color, they found their Pressiah, and then slowly returned to their lives as the camps disbanded.

If this all sounds frivolous, it should—here were millions of people bonded by a button that did nothing at all. The pull of feedback is so great that people will spend weeks online waiting to learn what will happen when they refrain from pushing a virtual button for sixty seconds.

In 1971, a psychologist named Michael Zeiler sat in his lab across from three hungry White Carneaux pigeons. The birds looked more like plump doves than common gray pigeons, and they were good eaters and quick learners. At the time, many psychologists were trying to understand how animals respond to different forms of feedback. Most of the work focused on pigeons and rats, because they were less complicated and more patient than humans, but the research program had lofty aims. Could the behavior of lower-order animals teach governments how to encourage charity and discourage crime? Could entrepreneurs inspire overworked shift workers to find new meaning at work? Could parents learn how to shape perfect children?

Before Zeiler could change the world, he had to work out the best way to deliver rewards. One option was to reward every desirable behavior, in the same way that some factory workers are rewarded for every gadget they assemble. Another was to reward those same desirable behaviors on an unpredictable schedule, creating some of the mystery that encourages people to buy lottery tickets. The pigeons had been raised in the lab, so they knew the drill. Each one waddled up to a small button and pecked persistently, hoping that it would release a tray of Purina pigeon pellets. The pigeons were hungry, so these pellets were like manna. During some trials, Zeiler would program the button so it delivered food every time the pigeons pecked; during others, he programmed the button so it delivered food only some of the time. Sometimes the pigeons would peck in vain, the button would turn red, and they'd receive nothing but frustration.

When I first learned about Zeiler's work, I expected the consistent schedule to work best. If the button doesn't predict the arrival of food perfectly, the pigeon's motivation to peck should decline, just as a factory worker's motivation would decline if you only paid him for some of the gadgets he assembled. But that's not what happened at all. Like tiny feathered gamblers, the pigeons pecked at the button more feverishly when it released food 50–70 percent of the time. (When Zeiler set the button to produce food only once in every ten pecks, the disheartened pigeons stopped responding altogether.) The results weren't even close: they pecked almost twice as often when the reward wasn't guaranteed. Their brains, it turned out, were releasing far more dopamine when the reward was unexpected than when it was predictable. Zeiler had documented an important fact about positive feedback: that less is often more. His pigeons were drawn to the mystery of mixed feedback just as humans are attracted to the uncertainty of gambling.

Thirty-seven years after Zeiler published his results, a team of Facebook web developers prepared to unleash a similar feedback experiment on hundreds of millions of humans. Facebook has the power to run human experiments on an unprecedented scale. The site already had two hundred million users at the time—a number that would triple over the next three years. The experiment took the form of a deceptively simple new feature called a "like" button. Anyone who has used Facebook knows how the button works: instead of wondering what other people think of your photos and status updates, you get real-time feedback as they click (or don't click) a little blue-and-white thumbs-up button beneath whatever you post. (Facebook has since introduced other feedback buttons, so you're able to communicate more complex emotions than simple liking.)

It's hard to exaggerate how much the "like" button changed the psychology of Facebook use. What had begun as a passive way to track your friends' lives was now deeply interactive, and with exactly the sort of unpredictable feedback that motivated Zeiler's pigeons. Users were gambling every time they shared a photo, web link, or status update. A post with zero likes wasn't just privately painful, but also a kind of public condemnation: either you didn't have enough online friends, or, worse still, your online friends weren't impressed. Like pigeons, we're more driven to seek feedback when it isn't guaranteed. Facebook was the first major social networking force to introduce the like button, but others now have similar functions. You can like and repost tweets on Twitter, pictures on Instagram, posts on Google+, columns on LinkedIn, and videos on YouTube.

The act of liking subsequently became the subject of etiquette debates. What did it mean to refrain from liking a friend's post? If you liked every third post, was that an implicit condemnation of the other posts? Liking became a form of basic social support—the online equivalent of laughing at a friend's joke in public. Likes became so valuable that they spawned a start-up called Lovematically. The app's founder, Rameet Chawla, posted this introduction on its homepage:

> It's our generation's crack cocaine. People are addicted. We experience withdrawals. We are so driven by this drug, getting just one hit elicits truly peculiar reactions.
>
> I'm talking about Likes.

They've inconspicuously emerged as the first digital drug to dominate our culture.

Lovematically was designed to automatically like every picture that rolled through its users' newsfeeds. If likes were digital crack, Lovematically's users were pushing the drug at the heavily discounted rate of free. It wasn't even necessary to impress them anymore; any old post was good enough to inspire a like. At first, for three experimental months, Chawla was the app's only user. During that time, he automatically liked every post in his feed, and, apart from enjoying the warm glow that comes from spreading good cheer, he also found that people reciprocated. They liked more of his photos, and he attracted an average of thirty new followers a day, a total of almost three thousand followers during the trial period. On Valentine's Day 2014, Chawla allowed five thousand Instagram users to download a beta version of the app. After only two hours, Instagram shut down Lovematically for violating the social network's Terms of Use.

"I knew way before launching it that it would get shut down by Instagram," Chawla said. "Using drug terminology, you know, Instagram is the dealer and I'm the new guy in the market giving away the drug for free." Chawla was surprised, though, that it happened so quickly. He'd hoped for at least a week of use, but Instagram pounced immediately.

───────

When I moved to the United States for grad school in 2004, online entertainment was limited. These were the days before Instagram, Twitter, and YouTube, and Facebook was limited to students at Harvard. I had a cheap Nokia phone that was in-destructible but primitive, so the web was tethered to my dorm room. One evening, after work, I stumbled on a game called Sign of the Zodiac (*Zodiac* for short) that demanded very little mental energy. Zodiac was a simple online slot machine much like the actual slot machines in casinos: you decided how much to wager, and then you lazily clicked a button over and over again and watched as the machine spat out wins and losses. At first I played to relieve the stress of long days filled with too much thinking, but the brief ding that followed each small win, and the longer melody that followed each major win, hooked me fast. Eventually screenshots of the game would intrude on my day. I'd picture five pink scorpions lining up for the game's highest jackpot, followed by the jackpot melody that I can still conjure today. I had a minor behavioral addiction, and these were the sensory hangovers of the random, unpredictable feedback that followed each win.

My Zodiac addiction wasn't unusual. For thirteen years Natasha Dow Schüll, a cultural anthropologist, studied gamblers and the machines that hook them. The following descriptions of slot machines come from gambling experts and current and for-mer addicts:

> Slots are the crack cocaine of gambling.
> They're electronic morphine.
> They're the most virulent strain of gambling in the history
>   of man.
> Slots are the premier addiction delivery device.

These are sensationalized descriptions, but they capture how easily people become hooked on slot machine gambling. I can relate, because I became addicted to a slots game that wasn't even doling out real money. The reinforcing sound of a win after the silence of several losses was enough for me.

In the United States, banks aren't allowed to handle online gambling winnings, which makes online gambling practically illegal. Very few companies are willing to fight the system, and the ones that do are quickly defeated. That sounds like a good thing, but free and legal games like Sign of the Zodiac are also dangerous. At casinos, the deck is stacked heavily against the player; on average the house has to win. But the house doesn't have to win in a game without money. As David Goldhill, the C.E.O. of the Game Show Network, which also produces many online games, told me, "Because we're not restricted by having to pay real winnings, we can pay out one hundred and twenty dollars for every hundred dollars played. No land-based casino could do that for more than a week without going out of business." As a result, the game can continue forever because the player never runs out of chips. I played Sign of the Zodiac for four years and rarely had to start a new game. I won roughly 95 percent of the time. The game only ended when I had to eat or sleep or attend class in the morning. And sometimes it didn't even end then.

In contrast to free games, casinos win most of the time—but they have a clever way of convincing gamblers that the outcomes are reversed. Early slot machines were incredibly simple devices: the player pulled the machine's arm (hence the term "one-armed bandit") to spin its three mechanical reels. If the center of the reels displayed two or more of the same symbol when they stopped spinning, the player won a certain number of coins or credits. Today, slot machines allow gamblers to play multiple lines, in some cases as many as several hundred at once. The machine below, for example, allows you to play fifteen lines:

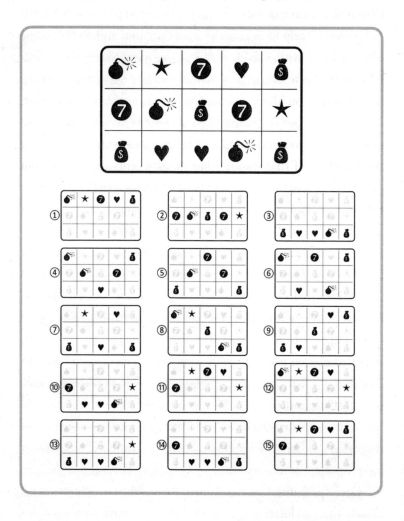

Say the machine charges ten cents per spin. If you decide to play all fifteen lines, each spin will cost you $1.50. Basically, you're playing fifteen spins at once, instead of drawing the experience out by playing a single spin fifteen times. Casinos are very happy for you to play this way: if they're going to beat you, they'll do it fifteen times more quickly. But every time you play, you're fifteen times more likely to win on at least one line, and the machine will celebrate with you by flashing the same bright lights and playing the same catchy tunes. Now imagine you play all fifteen lines, costing you $1.50, and one of your lines spins two bombs in a row, as line four does, above. If two bombs are worth a payout of ten credits, you get a payout of $1. Not bad—until you realize the net effect of that spin is a loss of fifty cents (your $1 payout minus the cost of the spin at $1.50). And yet you enjoy the positive feedback that follows a win—a type of win that Schüll and other gambling experts call a "loss disguised as a win."

Mike Dixon, a psychologist, has analyzed these disguised losses. With several colleagues, he focused on a game called Lucky Larry's Lobstermania (which I found online and played for three hours while I was supposed to be writing this book—I was lucky that U.S. laws forced me to play the free version). Lobstermania allows players to spin up to fifteen lines simultaneously. The game features five reels with three visible symbols per reel, for a total of more than 259 million possible outcomes. Dixon and his team calculated that gamblers are more likely to strike a loss disguised as a win than a genuine win any time they play six or more lines per spin.

Losses disguised as wins only matter because players don't classify them as losses—they classify them as wins. Dixon and his team hooked up a group of novice gamblers to electrodes while they played Lobstermania. He gave them ten dollars each, and told them they could win up to an additional twenty dollars.

They played for half an hour and spun, on average, 138 times. After each spin, a machine registered minute changes in how much the students were sweating—a sign that the event was emotionally meaningful. Lobstermania, like many modern video slots, is full of reinforcing feedback. In the background, the bouncy B-52s song "Rock Lobster" plays over and over whenever you spin. It's replaced by silence after losing spins and by louder, bouncier versions of the song after wins. Lights flash and bells ding just the same whether the spin represents a true win or a loss disguised as a win. The students sweated more when they won than when they lost—but they sweated just as much when their losses were disguised as wins as when those wins were genuine. This is what makes modern slot machines—and modern casinos—so dangerous. Like the little boy who hit every button in my elevator, adults never really grow out of the thrill of attractive lights and sounds. If our brains convince us that we're winning even when we're actually losing, how are we supposed to muster the self-control to stop playing?

After a string of losses, even die-hard gamblers begin to lose interest, some faster than others. This is a big problem for casinos, which aim to keep the gambler in front of the machine for as long as possible. It would be easy to change the odds of winning so that players become more and more likely to win after a series of losses, but, unfortunately for casinos, this is illegal in the U.S. The odds need to stay consistent across every spin, regardless of the previous run of outcomes. Natasha Dow Schüll told me that casinos have come up with some creative solutions. "Many casinos use 'luck ambassadors.' They sense that you're reaching your pain point—the moment when you're about to leave the casino—

and they dispatch someone to give you a bonus." These bonuses were either meal vouchers or a free drink or even cash or gambling credits. Bonuses are classified as "marketing" rather than a way of changing the odds of winning, so regulators turned a blind eye. With a new dose of positive reinforcement, gamblers tended to continue playing anew, until they reached another pain point after a series of losses.

It's expensive, however, to keep dozens of luck ambassadors on the floor, not to mention paying a team of data analysts to identify frustrated gamblers. One man, a casino consultant named John Acres, proposed a creative solution that skirted the relevant laws. Schüll explained Acres's technique. "As you play, a tiny portion of what you lose goes into a pot which counts as the marketing bonus pot. An algorithm within the machine senses your pain points, and knows ahead of time what the next outcome will be." Normally the algorithm sits by and lets the machine dish out a randomly drawn outcome. When the player reaches a pain point, though, it intervenes. "If the machine sees that, oh, that outcome sucks," Schüll said, "instead of BAR, BAR, CHERRY, it goes 'chink' and nudges the third reel so that it displays BAR—a jackpot outcome of three BARs." Those winnings are taken from the "marketing bonus pot" that grew in size while the player continued to lose. Instead of relying on a human luck ambassador, the machine plays that role itself. Schüll has seen many dastardly tactics in her time investigating casinos, but this one she calls "shocking." When she asked Acres how this wasn't "a complete violation of laws in place to protect people from precisely this," he replied, "Well, laws are made to be broken."

The success of slot machines is measured by "time on device." The longer the average player stays seated at the machine, the better the machine. Since most players lose more money the longer they play, time on device is a useful proxy for profitability. Video game designers use a similar measure, which captures how engaging and enjoyable their games are. The difference between casinos and video games is that many designers are more concerned with making their games fun than with making buckets of money. Bennett Foddy, who teaches game design at New York University's Game Center, has created a string of successful free-to-play games, but each was a labor of love rather than a moneymaking vehicle. They're all available on his website, foddy. net, and apart from attracting limited advertising revenue, they aren't a significant source of income, despite some having achieved cult status.

"Video games are governed by microscopic rules," Foddy says. "When your mouse cursor moves over a particular box, text will pop up, or a sound will play. Designers use this sort of micro-feedback to keep players more engaged and more hooked in." A game must obey these microscopic rules, because gamers are likely to stop playing a game that doesn't deliver a steady dose of small rewards that make sense given the game's rules. Those rewards can be as subtle as a "ding" sound or a white flash whenever a character moves over a particular square. "Those bits of micro-feedback need to follow the act almost immediately, because if there's a tight pairing in time between when I act and when something happens, then I'll think I was causing it." Like kids who push elevator buttons to see them light up, gamers are motivated by the sense that they're having an effect on the world. Remove that and you'll lose them.

The game Candy Crush Saga is a prime example. At its peak in 2013, the game generated more than $600,000 in revenue per day. To date, its developer, King, has earned around $2.5 billion from the game. Somewhere between half a billion and a billion people have downloaded Candy Crush Saga on their smartphones or through Facebook. Most of those players are women, which is unusual for a blockbuster. It's hard to understand the game's colossal success when you see how straightforward it is. Players aim to create lines of three or more of the same candy by swiping candies left, right, up, and down. Candies are "crushed"— they disappear—when you form these matching lines, and the candies above them drop down to take their place. The game ends when the screen fills with candies that can't be matched. Foddy told me that it wasn't the rules that made the game a success—it was *juice*.

Juice refers to the layer of surface feedback that sits above the game's rules. It isn't essential to the game, but it's essential to the game's success. Without juice, the same game loses its charm. Think of candies replaced by gray bricks and none of the reinforcing sights and sounds that make the game fun. "Novice game designers often forget to add juice," Foddy said. "If a character in your game runs through the grass, the grass should bend as he runs through it. It tells you that the grass is real and that the character and grass are in the same world." When you form a line in Candy Crush Saga, a reinforcing sound plays, the score associated with that line flashes brightly, and sometimes you hear words of praise intoned by a hidden, deep-voiced Wizard of Oz narrator.

Juice is effective in part because it triggers very primitive parts of the brain. To show this, Michael Barrus and Catharine Winstanley, psychologists at the University of British Columbia, created a "rat casino." The rats in the experiment gambled for delicious sugar pellets by pushing their noses through one of four small holes. Some of the holes were low-risk options with small rewards. One, for example, produced one sugar pellet 90 percent of the time, but punished the rat 10 percent of the time by forcing him to wait five seconds before the casino would respond to his next nose poke. (Rats are impatient, so even small waits register as punishments.) Other holes were high-risk options with larger rewards. The riskiest hole produced four pellets, but only 40 percent of the time—on 60 percent of trials, the rat was forced to wait in time-out for forty seconds, a relative eternity.

Most of the time, rats tend to be risk-averse, preferring the low-risk options with small payouts. But that approach changed completely for rats who played in a casino with rewarding tones and flashing lights. Those rats were far more risk-seeking, spurred on by the double-promise of sugar pellets and reinforcing signals. Like human gamblers, they were sucked in by juice. "I was surprised, not that it worked, but how well it worked," Barrus said. "We expected that adding these stimulating cues would have an effect. But we didn't realize that it would shift decision making so much."

Juice amplifies feedback, but it's also designed to unite the real world and the gaming world. One of Foddy's most successful games is called Little Master Cricket, which does this very well. In the game, a cricket player hits one shot after another, scoring runs (or points) according to where those shots go. When he misses the ball or hits it in the wrong spot, he's "out" and the game begins again at zero runs. "When I released Little Master, my wife was working at the head offices of Prada in New York," Foddy said. "Much of the finance department consisted of cricket fans from India—and they were hooked." When they discovered that their colleague was married to the game's creator, they were starstruck. It's very difficult to simulate the game of cricket in an engaging way, but Foddy somehow managed to keep the game simple but true to life. Players move the mouse back and forth in a way that mirrors the swing of a real cricket batsman. Just as in real life, the highest scoring shots in Little Master travel far through the air while avoiding the clutches of fielders who might catch the ball before it falls to the ground. (As in baseball, this renders the batsman "out.") This sort of feedback, which ties the game to the real world, is called mapping. "Mapping is sort of visceral," says Foddy. "For example, you should always use the space bar sparingly. It's a loud, clattery key on the computer, so it shouldn't be used for something mundane, like walking. It's better saved for declarative actions that aren't quite as common, like jumping. Your aim is to match sensations in the physical realm to those in the digital realm."

The most powerful vehicle for juice must surely be virtual reality (VR) technology, which is still in its infancy. VR places the user in an immersive environment that can be real (a beach on the other side of the world) or imaginary (the surface of Mars).

The user navigates and interacts with that world as she might the real world. Advanced VR also introduces multisensory feedback, including touch, hearing, and smell.

In a podcast released on April 28, 2016, author and sports columnist Bill Simmons asked billionaire investor Chris Sacca about his experience with VR. "I'm afraid for my kids, a little bit," Simmons told Sacca. "I do wonder if this VR world you dive into is almost superior to the actual world you're in. Instead of having human interactions, I can just go into this VR world and do VR things and that's gonna be my life." Sacca, an early Google employee and Twitter investor, shared Simmons' concerns:

> That's very legit. One of the things that's interesting about technology is that the improvement in resolution and sound modeling and responsiveness is outpacing our own physio-logical development. Our biology has been the same—we weren't built to ingest all this light and sound in this incred-ibly coordinated way . . . you can watch some early videos . . . where you are on top of a skyscraper, and your body will not let you step forward. Your body is convinced that that is the side of the skyscraper. That's not even a super high-res or super immersive VR platform. So we have some crazy days ahead of us.

VR has been around for decades, but it's now on the cusp of going mainstream. In 2013, a VR company called Oculus VR raised $2.5 million on Kickstarter. Oculus VR was promoting a headset for video games called the Rift. Until recently, most people thought of VR as a tool for gaming, but that changed when Facebook acquired Oculus VR for $2 billion in 2014. Facebook's Mark Zuckerberg had big ideas for the Oculus Rift that went far beyond games. "This is just the start," Zuckerberg said. "After games, we're going to make Oculus a platform for many other experiences. Imagine enjoying a court side seat at a game, studying in a classroom of students and teachers all over the world or consulting with a doctor face-to-face—just by putting goggles in your home." VR no longer dwelled on the fringes. "One day, we believe this kind of immersive, augmented reality will become a part of daily life for billions of people," said Zuckerberg.

In October 2015, the *New York Times* shipped a small cardboard VR viewer with its Sunday paper. Paired with a smartphone, the Google Cardboard viewer streamed exclusive *Times* VR content, including documentaries on North Korea, Syrian refugees, and a vigil following the Paris terror attacks. I spent much of that Sunday afternoon lost in a documentary about child refugees, forgetting for long stretches of time that I wasn't actually standing in a devastated schoolroom in war-torn Ukraine. "Instead of sitting through forty-five seconds on the news of someone walking around and explaining how terrible it is, you are actively becoming a participant in the story that you are viewing," said Christian Stephen, a producer of one of the VR documentaries.

But the Google Cardboard pales next to the Oculus Rift. According to Palmer Luckey, founder of Oculus VR, "Google Cardboard is muddy water compared with the fancy wine of Oculus Rift." Of course, for the moment, Google Cardboard has the advantage of costing around $10 online, while the Oculus Rift sells for $599.

Despite the promise of VR, it also poses great risks. Jeremy Bailenson, a professor of communication at Stanford's Virtual Reality Interaction Lab, worries that the Oculus Rift will damage how people interact with the world. "Am I terrified of the world where anyone can create really horrible experiences? Yes, it does worry me. I worry what happens when a violent video game feels like murder. And when pornography feels like sex. How does that change the way humans interact, function as a society?"

In an article for the *Guardian*, tech writer Stuart Dredge noted that we're already struggling to focus our attention on friends and family. If idle smartphones and tablets draw us away from real-world interactions, how will we fare in the face of VR devices? Steven Kotler wrote for *Forbes* that VR would become "legal heroin; our next hard drug." There's every reason to believe Kotler. When it matures, VR will allow us to spend time with anyone in any location doing whatever we like for as long as we like. That sort of boundless pleasure sounds wonderful, but it has the capacity to render face-to-face interactions obsolete. Why live in the real world with real, flawed people when you can live in a perfect world that feels just as real?

Since mainstream VR is in its infancy, we can't be sure that it will dramatically change how we live. But all early signs suggest that it will be both miraculous and dangerous. As Zuckerberg said, it will allow us to see doctors who are thousands of miles away, to visit and learn about distant places (both inaccessible and imaginary) that we might never experience firsthand, and to "visit" loved ones who live across the world. Wielded by big business and game designers, though, it might also prove to be a vehicle for the latest in a series of escalating behavioral addictions.

In contrast to VR, the physical realm is a long series of losses punctuated by occasional wins. Gamers have to lose from time to time. A game that pays out all the time is no fun at all. When I met with David Goldhill, the C.E.O. of the Game Show Network, he told me a story that illustrates the surprising downsides of winning all the time. Goldhill is a natural storyteller. He radiates competence and reveals an uncanny command of any topic that comes up in conversation. We discussed my hometown, Sydney, and by the end of the conversation I was scribbling notes like a tourist. Goldhill's story involved a gambler who wins all the time. "The guy thinks he's in heaven because he wins every single bet. Eventually, though, he realizes that he's in hell. It's absolute torture." The gambler's been chasing wins all his life, and now that they're arriving one after another his reason for existing is gone. Goldhill's story illustrates why variable reinforcement is so powerful. Not because of the occasional wins, but because the experience of coming off a recent loss is deeply motivating.

The best part of any gamble may be the millisecond before the outcome reveals itself. This is the moment of maximum tension, when gamblers are primed to see a winning outcome. We know this from a clever experiment that two psychologists published in 2006. Emily Balcetis and Dave Dunning told a group of Cornell undergrads that they were participating in a juice taste test. Some of them would be lucky enough to try freshly squeezed orange juice, but others would drink a "gelatinous, chunky, green, foul-smelling, somewhat viscous concoction labeled as an 'organic veggie smoothie.'" As the students inspected each beverage, the experimenter explained that a computer would randomly assign them to drink a tall glass of one or the other. Half the students were told that the computer would present a number if they were assigned to drink the appealing orange juice (and a letter if they were assigned to drink the sludge), while the other half were told the reverse, that the letter spelled salvation and the number spelled doom. The students sat at the computer and waited, a lot like the gamblers waiting for a slot machine to display its outcome. A couple of seconds later the computer displayed this figure:

Eighty-six percent of them rejoiced. The computer had come through with a win!

As you've probably gathered, the figure is neither a number nor a letter, but instead an ambiguous hybrid of the number 13 and a capital letter B. The students were so intent on seeing what they hoped to see that their brains resolved the ambiguous figure in their favor. The number thirteen popped out to those who hoped to see a number, and the letter B popped out to those who hoped to see a letter. This phenomenon, called motivated perception, happens automatically all the time. It's usually hidden to us, but Balcetis and Dunning were clever enough to find a way to unmask the effect.

What makes motivated perception so important for addiction is that it shapes how we perceive negative feedback. David Goldhill's story shows us that gamblers hate to win all the time—but even more than that, they hate losing all the time. If hapless gamblers and gamers and Instagram users saw the world as it really is, they'd see that they lose most of the time. They'd recognize that a string of losses usually foretells more losses, rather than an approaching jackpot, and that the figure above is just as likely to be a letter as it is a number. To make matters worse, many games and gambling experiences are designed to get your hopes up by displaying near wins. In a classic early episode of *The Simpsons* from Season One, Homer Simpson buys a scratch card lottery ticket from Apu at the Kwik-E-Mart:

Homer: One glazed, and one Scratch-'N-Win, please.
[Apu hands Homer his lottery ticket and he starts to scratch it off.]
Homer: Oh. Liberty Bell.
[Homer scratches some more and gasps.]
Homer: Another Liberty Bell! One more and I'm a million-

aire. Come on, Liberty Bell, please, please, please, please, please, please!

[Homer scratches to reveal a plum.]

Homer: D'oh! That purple fruit thing. Where were you yesterday?

Homer's disappointment is shared by millions of scratch card near winners every week. Yesterday Homer "almost won" with two "purple fruit things" and today he almost won with two Liberty Bells. There's a pretty good chance he'll play again tomorrow and the next day, because to Homer this wasn't a loss. It was an "almost win."

6.

Progress

Shigeru Miyamoto knows how to design a video game that people can't stop playing. He is the gaming world's answer to Steven Spielberg or Stephen King or Steve Jobs—an artist who understands what people want better than they do, and who turns everything he touches to gold. Miyamoto was behind the second-highest grossing game of all time. And also the games ranked fifth, sixth, eighth, ninth, eleventh, twelfth, nineteenth, twenty-first, twenty-third, twenty-fifth, twenty-sixth, thirty-third, and thirty-fourth. The industry would have been much the poorer without his influence. What Miyamoto seemed to recognize better than anyone was that addictive games offered something to both novices and experts. Games designed only for beginners would grow stale too soon, and games designed only for experts would lose newcomers before they became masters.

When Miyamoto was twenty-four he joined Nintendo. For ninety years Nintendo had traded in the stagnant playing card business, but now, in the late 1970s, it was branching out into video games. As a young man Miyamoto had fallen in love with the arcade game Space Invaders, so his father pulled some strings to arrange an interview for his son with Nintendo's president. Miyamoto showed the president some of the toys and games he'd created in his spare time, and was hired on the spot as an apprentice video game planner.

The early 1980s were difficult for Nintendo. The company tried to generate a U.S. market for video games, but failed dismally. Thousands of unsold games were languishing in a warehouse when Nintendo's head engineer approached young Miyamoto and asked him to design a new game that would save the dying company. According to the ever modest Miyamoto, "no one else was available to do the work." Miyamoto's first game was a classic named Donkey Kong. The young hero of the game was a mustachioed plumber named Mario, who was named for Nintendo America's warehouse landlord, Mario Segale. The same Mario would go on to feature in one of the best-selling series of all time, Super Mario Bros. Super Mario was where Miyamoto showcased his ability to make games attractive to players at all levels.

Super Mario Bros. hooks newcomers because there are no barriers to playing the game. You can know absolutely nothing about the Nintendo console and still enjoy yourself from the very first minute. There's no need to read motivation-sapping manuals or grind through educational tutorials before you begin. Instead, your avatar, Mario, appears on the left-hand side of an almost empty screen. Because the screen is empty, you can push the Nintendo controller's buttons randomly and harmlessly, learning which ones make Mario jump and which ones make him move left and right. You can't move any further left, so you quickly learn to move right. And you aren't reading a guide that tells you which keys are which—instead, you're learning by doing, and enjoying the sense of mastery that comes from acquiring knowledge through experience. The first few seconds of gameplay are brilliantly designed to simultaneously do two very difficult things: teach, and preserve the illusion that nothing is being taught at all.

L ike thousands of kids I fell hard for Super Mario Bros. I was ten years old and my family was visiting relatives in New Zealand. My aunt introduced me to a boy who was my age, and, as ten-year-old boys tend to, he showed me his action figures and his Nintendo. I'd never seen a Nintendo before, and when he fired up Super Mario Bros., he unintentionally ruined the rest of my vacation. We played for half an hour, but by the end of that visit the game was all I could think about.

Decades later and thousands of miles from New Zealand, I gave a lecture that was inspired by Miyamoto and a Yale economist named Martin Shubik. The two men were from different worlds, but both had devised traps that took hold almost immediately. Shubik described his trap in a journal article that he published in 1971: "There is an extremely simple, highly amusing, and instructive parlor game which can be played at any party by arranging for the auction of a dollar." Shubik described the rules

of his so-called Dollar Auction Game as "simplicity itself." Here
they are in their entirety:

> The auctioneer auctions off a dollar bill to the highest bid-
> der, with the understanding that *both* the highest bidder and
> the second highest bidder will pay.

If one person is willing to pay eighty cents for the dollar bill,
and another person, the second-highest bidder, is willing to pay
seventy cents, the auctioneer takes in one dollar and fifty cents—
a tidy profit of fifty cents. Both bidders pay, but only the highest
bidder gets the dollar bill. This is a great deal, obviously, because
she's paying eighty cents for a bill that's literally worth a dollar.
For the second-highest bidder, though, it's a terrible deal. He pays
seventy cents for precisely nothing.

I played Shubik's game in my lecture, but I auctioned off a
twenty-dollar bill. Bids started at one dollar, and rose in incre-
ments of one dollar. A dozen voices immediately shouted "one
dollar!" because paying a dollar for a twenty-dollar bill is a great
investment. I heard "two dollars!" and then "three dollars!" Some
of the students stopped bidding early on, but others continued
past ten dollars, on toward the magic twenty-dollar mark. When
you watch people taking part, you can see on their faces the exact
moment when they realize that the game is a trap. When the
number of active bidders inevitably drops to just two, one of those
people has to pay for absolutely nothing. For example:

> Person A: Sixteen dollars!
> Person B: Seventeen dollars!

. . . pause . . .

Person A: Eighteen dollars!

Person B: Nineteen dollars!

Were this a normal auction, the game would end here. There's no reason for Person A to shout out "Twenty dollars!" unless he really doesn't like Person B and would rather make zero profit (paying twenty dollars for a twenty-dollar bill) than watch Person B earn a dollar.

But this is a trap, and so the bidding escalates:

Person A: Twenty dollars!

. . . pause . . .

Person B: Twenty-one dollars!

. . . longer pause . . .

Person A: Twenty-two dollars!

. . . even longer pause . . .

Person B (more quietly): Twenty-three dollars.

Sometimes the game goes on to triple or even quadruple the bill's worth. No one wants to pay a huge sum of money for nothing, which makes the Twenty Dollar Auction Game a terrific way to raise money for charity.

Shubik's game shows that an early hook fuels many addictive behaviors. The experience seems innocuous at first, but eventually you realize that things might end badly. For my students, the hook was the slim chance of winning twenty dollars at a heavy discount. In my case, the hook was a plumber named Mario in search of a kidnapped princess.

The Dollar Auction Game hooks beginners fast, but it also works so effectively because it functions a bit like a bait-and-switch campaign. Bait-and-switch campaigns are the sort of illegal advertisements that electrical goods retailers use to lure Christmas sale shoppers. A store might advertise, for example, a new DVD player—"$9 while stocks last!"—but only keep one of those DVD players in stock. Customers line up around the block, storm into the store at 9 A.M., and one goes home with the DVD player while fifty are left with a horrible choice. Psychologically speaking, they already *feel* like owners of an inexpensive new DVD player. While braving the cold in line two hours earlier, they started to imagine what it would be like to watch the eight Harry Potter films with the whole family gathered around a huge bowl of popcorn. The choice, then, is to abandon those fantasies, or pay $199 for the next cheapest DVD player now that the $9 model has vanished.

This is what the Dollar Auction Game does, too. Bidders form an emotional attachment to winning the auction. For the two students bidding up to $60 in my classroom, the motivation isn't the thrill of winning $20—it's the threat of losing to the other bidder. As neuroscientist Kent Berridge suggested, their facial expressions show that they *want* to keep bidding, but they're certainly not *liking* the experience at all.

You can see the same loss aversion even more clearly in so-called penny auction websites like Quibids.com, HappyBidDay .com, and Beezid.com. To begin using Beezid, for example, you buy a pack of bids. Packs range in size from forty bids (for $36, or

90 cents per bid) to one thousand bids (for $550, or 55 cents per bid). The Beezid site features hundreds of ongoing auctions for products like laptops, TVs, and headphones. This is how an auction for a new TV looks after the first bid:

Sigmasonic 50-inch class4K LED TV

Compare to: 2,999.00

5:00:00

Get it for:

0.01

bidking999

BID

The first bid was for one cent—a single cent!—and it was placed by a user named bidking999. The clock reads five hours, which means that bidking999 will win the TV for the princely sum of one cent if no one else bids before five hours have elapsed. Each bid usually raises the price of the item by one cent (hence the term *penny auction*). Bidding is scattered at first, but when the clock drops below roughly fifteen seconds, the auction enters "action time," during which every new bid restarts the clock at fifteen seconds. On particularly hot items, this happens dozens of times—a bit like the Reddit April Fools' countdown button that took weeks to reach zero. Some items sell very cheaply, but others sell at close to face value. The problem for the consumer is having to bid thousands of times before winning anything, which burns through thousands of pre-purchased bids while paying nothing in return. The site makes a tidy profit, while the consumer gradually loses a few cents at a time until his losses become enormous.

Hundreds of penny auction participants complain online. Some argue that the site is a scam, and others compare it to gambling. One consumer reports expert at SiteJabber.com tested a penny auction platform and said that, despite being wary, "Even I was drawn to those sites and I felt like I was putting quarters into a slot machine but with no real chance of winning." The process is so addictive because you pay for bids up front, so spending them doesn't feel painful at all, and the lure of saving thousands of dollars—in this case paying one penny for a three-thousand-dollar TV—is hard to resist. Once the bidding period enters action time, you can almost taste the win. The stakes are low when you place your first bid; but by the time you've placed your hundredth bid, and you've seen the timer drop to one second dozens of times, you're heavily invested in the process. No wonder consumer reports sites call penny auction platforms "risky," classify them as "scams," and often recommend buyers avoid them altogether.

Penny auction sites have earned their awful reputations, but not every engaging experience is predatory. Some experiences are designed to be addictive for the sake of ensnaring hapless consumers, but others happen to be addictive though they're primarily designed to be fun or engaging. The line that separates these is very thin; to a large extent the difference rests on the intention of the designer. Penny auction sites are predatory by design, as are slot machines. (Natasha Dow Schüll titled her book on gambling *Addiction by Design*.) But when Shigeru Miyamoto designed Super Mario Bros., his primary aim was to make a game that he himself enjoyed playing. Instead of consulting focus groups, he played the game for hours on end, ironing out its bugs and settling, in time, on the version that Nintendo released in 1983. In the 1990s and early 2000, Miyamoto designed the wildly successful Pokémon games, and again his primary allegiance was to the integrity of the game. "That's the point," he said, "not to make something sell, something very popular, but to love something, and make something that we creators can love. It's the very core feeling we should have in making games." When you compare Super Mario Bros.—regularly voted by game designers as the greatest game of all time—to others on the market, it's easy to recognize in the competition the hallmarks of a predatory game.

Adam Saltsman, who produced an acclaimed indie game called Canabalt in 2009, has written extensively about the ethics of game design. "Predatory games are designed to abuse the way you're wired," Saltsman said. "Many of the predatory games of the past five years use what's known as an energy system. You're allowed to play the game for five minutes, and then you artificially run out of stuff to do. The game will send you an email in, say, four hours when you can start playing again." I told Saltsman that the system sounded pretty good to me—it forces gamers to take breaks and encourages kids to do their homework between gaming sessions. But that's where the predatory part comes in. According to Saltsman, "Game designers began to realize that players would pay one dollar to shorten the wait time, or to increase the amount of energy their avatar would have once the four-hour rest period had passed." The game ensnares you, like penny auctions and Shubik's Dollar Auction Game do, and manipulates you into waiting or paying. I came across this predatory device when playing a game called Trivia Crack. If you give the wrong answer several times, you run out of lives, and a dialogue screen gives you a choice: wait for an hour for more lives, or pay ninety-nine cents to continue immediately.

Many games hide these down-the-line charges. They're free, at first, but later you're forced to pay in-game fees to continue. "Those hidden charges are one way of behaving disrespectfully toward your audience of players," Saltsman said. "They're a bit like the classic arcade games that charged you a quarter to play the easy opening level, but then forced you to confront a really tough boss at the end of the level. The whole level is easy and fun to play, and then the boss is super hard to defeat. So you have to put in lots of extra quarters to get to the next fun level. The game advertises itself as costing a quarter, but there's no way to kill the boss without spending a dollar or more." If you're minutes or even hours deep into the game, the last thing you want to do is admit defeat. You have so much to lose, and your aversion to that sense of loss compels you to feed the machine *just one more time*, over and over again. You start playing because you want to have fun, but you continue playing because you want to avoid feeling unhappy.

Even if the industry's biggest game designers aren't sure how to make their games addictive, they learn quickly with the help of a clever technique. "It's called color coding," Isaac Vaisberg, the former gaming addict I introduced in chapter 2, told me. He gave the example of an online role-playing game, in which players form guilds to complete missions. "Say you have two million players already, and you're trying to figure out what's most engaging to them. You attach a color to the [computer] code associated with each mission, or even to different elements within each mission, and see which is most addictive." The color codes, or tags, allow designers to track how much time players spend on each element within each mission, and how many times they come back to try the mission again. "Since you have a huge sample of players, you can run experiments. Mission A might require you to save something, whereas Mission B is very similar except that you have to kill something." Similarly, Mission C might give you a burst of positive feedback early on, while Mission D, which is otherwise identical, doesn't give you any feedback. A designer can see that, for example, people spend three times as long playing a mission that requires them to kill rather than save, and return 50 percent more often to a mission that gives them short bursts of micro-feedback. The result is a weaponized version of the original game that evolves over time to be maximally addictive. "World of Warcraft is particularly good at this," Vaisberg said about the game that ensnared him for a couple of years.

"Over eight years they've engineered the game to include every-thing people like." FarmVille, for instance, is an addictive game about running a virtual farm. At its peak, tens of millions of Facebook users played the game. "FarmVille was huge on Face-book, especially among women, so the World of Warcraft team embedded a version of FarmVille within World of Warcraft to attract female gamers."

Historically most gamers have been men, but the gaming world has begun to appeal to women and other underserved groups. In fact, in August 2014, women over the age of eighteen became the largest demographic in gaming. They represent 36 percent of gamers, whereas men over the age of eighteen make up 35 percent of all gamers. This rise was fueled, in part, by games like Kim Kardashian's Hollywood. Kardashian released the game in June 2014, and in its first year it took in tens of mil-lions of dollars. Almost half of the game's revenue went to Kar-dashian herself. The game is free to download, but there's a tiny "in-app purchases" warning under the download button, and it's almost impossible to play the game without spending money. The aim is to rise from E-list to A-list celebrity by doing the sorts of things that Kardashian herself might do: change your outfit often, be seen in public, parade your friends wherever you go, date lots of people, and, above all, avoid getting dumped. Players earn K-stars whenever their celebrity stars rise, but to make meaning-ful progress they have to buy booster packs. A small star pack is five dollars, but an extra-large pack costs forty dollars. You can also spend your real, hard-earned cash to buy virtual cash.

Like World of Warcraft, Kardashian's game delivers small doses of positive feedback to entice players as soon as they begin.

The game's production company, Glu Games, does plenty of testing to make sure those rewards are delivered at precisely the right intervals. One *Business Insider* columnist declared the game "uniquely toxic and addictive . . . perhaps the only app that really deserves the comparison to drugs." Other journalists reported similar addictions. *Jezebel*'s Tracie Morrissey admitted spending nearly five hundred dollars on the game: "You guys, I literally think I have a problem. What a lame, embarrassing addiction to have. What would I even say if I tried to get help for this at AA or something?" Emilee Lindner wrote an article on MTV.com titled "True Life: I Got Addicted to the Kim Kardashian Game," and admitted using most of her family's data plan when she played, sometimes through the night. Many of these "addicts" are high-functioning people who otherwise hold down impressive jobs and raise families. They aren't the stereotypical addicts of yesteryear, which is precisely what makes the products that seize them so insidious. One minute, they're novices passing time with a new, free game, and the next they're apologizing for blowing the family budget on gameplay.

In the world of budding addictions, beginner's luck is a serious hazard. When I was eight and my brother was six, we visited the local ten-pin bowling alley with our parents for the first time. Bowling is a difficult game for adults and a terrible game for children. Modern alleys deal with this by replacing the gutters alongside the bowling lane with bumpers, which make it impossible to bowl a gutter ball. A game of skill becomes a game of luck as the ball bounces wildly off the bumpers. In the late 1980s, when we visited, there were no bumpers, and there were no concessions for hopeless beginners.

We paid for our two games and walked past an endless row of bowling balls. The first rack was filled with a dozen black sixteen-pound balls. These were the no-nonsense bowling balls reserved for serious bowlers—strong men with big hands who could turn their wrists to impart enormous spin on the game's heaviest ball. We walked past the fifteen-pounders, fourteen-pounders, and down to the far end of the alley where a small rack held several balls made for younger bowlers. You could tell they were for kids because they were pink and blue and orange, and because the finger holes were almost too small even for our tiny paws. Also, they weighed six pounds.

We broke no records that day, but my brother was forever hooked on the game. Where he enjoyed a freakish dose of beginner's luck, I bowled with consistent incompetence. I racked up a couple of pins here and there, and we finished on a similar score, but his eight points—his entire haul for the day—came on his first attempt. I still remember his shuffling approach, and the awkward double-handed sidearm throw that drove the ball much harder downward into the floor than it did toward the pins. By some miracle, the ball ambled down the lane, avoided the gutters, and very slowly toppled all but two of the pins. We cheered and he celebrated, but that was the last time he would score that day. For years afterward he was obsessed with the game, and I'm convinced that his obsession was driven in part by the early taste of success that came before a long era of failure.

Beginner's luck is addictive because it shows you the pleasure of success and then yanks it away. It gives you unrealistic ambitions and the high expectations of a more seasoned competitor. Your second dose of success is a mirage that seems nearer than it actually is, and the sense of loss that mounts with each new failure drives you ever harder till you recapture that early (and undeserved) sense of glory.

I watched my brother bowl a long string of gutter balls—not just that day but each time we went bowling for many years. More than twenty years later, I decided, with my colleagues Heather Kappes, Dave Berri, and Griffin Edwards, to replicate his experience in the lab. We invited a group of adults into a lab to play darts. None of them had played before. We told them their performance would be scored, but to be fair we gave them a chance to practice first. Half of them stood so close to the dartboard that success was all but guaranteed, while the other half practiced from much further back and generally struggled—a far more realistic piece of feedback. Later, when we asked everyone how much they enjoyed playing and how motivated they were to play again, the "lucky" beginners were game to continue. The unlucky beginners weren't completely discouraged, but their early dose of realistic feedback dampened their enthusiasm for the game.

Many game designers know that beginner's luck is a powerful hook. Nick Yee, who has a doctorate in communication and studies how games affect players, has written about the role of early rewards in online role-playing games.

One of [the factors that attract people to online role-playing games] is the elaborate rewards cycle inherent in them that works like a carrot on a stick. Rewards are given very quickly in the beginning of the game. You kill a creature with 2–3 hits. You gain a level in 5–10 minutes. And you can gain crafting skill with very little failure. But the intervals between these rewards grow exponentially fairly quickly. Very soon, it takes 5 hours and then 20 hours of game time before you can gain a level. The game works by giving you instantaneous gratification upfront and leading you down a slippery slope.

Designers discovered this tactic after combing millions of data points—the sort of exploratory exercise that Isaac Vaisberg described. Where my brother's beginner's luck was a true fluke, the "luck" that graces novice gamers is engineered.

_____

Beginner's luck is addictive, but some experiences are so friendly to beginners that luck is unnecessary. When I visited David Goldhill, the Game Show Network C.E.O. I mentioned earlier, he began by handing me his phone. "I want to show you a game that I find fascinating. My youngest kid who's seven years old loves it. It's incredibly simple and stupid. Do you know Crossy Road?" I told him I didn't. "See how long it takes you to figure out how to play the game." It took me three seconds. All your avatar has to do is cross the road without getting run over. He moves with simple taps of the screen. This "simple and stupid" game, like Super Mario Bros., is designed so there are no barriers to entry. The minute you see the screen, you know as much as you need to know to start making progress. "This reminds me of another game . . ." I told Goldhill before he interrupted me: "It reminds everyone of some other game they've played." Crossy

Road borrows elements from so many games that, if you've played just one or two of them, you've effectively played them all.

The Game Show Network hosts and produces games, but the organization is best known for its TV game shows. They work on the same principle. "If you watch a good game show that you've never seen before, within a couple of minutes of tuning in the rules will either be clear to you, or they'll actually be explained to you," Goldhill said. "Part of the design of a good game show is that there are no barriers to entry. And there's a worldwide vernacular. No matter where you are, if you tune in to a game show, they share the same set of basic elements. If you look on YouTube, you'll see fifteen-, sixteen-, seventeen-year-olds designing their own game shows, and they use that same vernacular."

I thought back to the games that had recently occupied my time and attention. Almost without exception, they were remarkably simple. Earlier I mentioned Adam Saltsman's game Canabalt, which is a perfect example. Your aim is to control a man who's running from some ambiguous alien threat along a futuristic cityscape, hopping from building to building, moving faster and faster as he goes. The game determines his running speed, so all you have to do is tap the screen when you want your avatar to jump. During a particularly turbulent flight across the Atlantic, I soothed my nerves by playing the game over and over again. It was the game's simplicity that made it the perfect vehicle for meditation. I know I must have looked odd, because I saw a friend play Canabalt once. His face was screwed up in concentration, his body completely motionless except for his index finger, which waggled cartoonishly up and down as he coaxed his avatar to jump—slowly at first, and then faster as the game wore on. There is no end to the game—you can play forever if you're superhuman—and Saltsman was credited with spawning a new genre of games called "endless runners." In a *New Yorker* interview, game designer Luke Muscat recalled, "I remember playing Canabalt and just thinking, How has nobody ever thought of this before?" As if to underscore the game's simplicity, Saltsman came up with the game's odd name by listening to his six-year-old nephew merging with the words "cannonball" and "catapult."

For decades, video games were played by teenage boys and men who never grew up. That's no longer true, because gamers don't need consoles or big chunks of free time. Smartphones have changed the gaming landscape completely. Take FarmVille, the game that WoW embedded in its platform. "FarmVille was wildly popular," says Frank Lantz, director of the New York University's Game Center. Roughly one in ten Americans have played FarmVille, and for two years it was the most popular game on Facebook. Players were charged with building a farm by tending to virtual crops and animals. The game was addictive and predatory: once players built their farms, they had to return to the game at preset intervals to water their crops. If the crops died, which happened to millions of players whose lives and sleep schedules sometimes prevented them from returning to the game, they could pay to "unwither" those crops. People spent untold sums of money undoing that neglect. *Time* called the game one of the fifty worst inventions of all time because its "series of mindless chores" was so addictive. "Harvest Moon was very similar to FarmVille," Lantz said, "but you had to own a Super Nintendo console to play the game. Well, these people playing FarmVille don't need a console, and it doesn't make sense for them to crouch down in front of a television set and play Harvest Moon. But here's a game you can play for five minutes at work, or whenever you want to take a break. In some ways it's very similar to an existing genre, but with a new rhythm that fits into these people's lives. It introduced people who hadn't played games before and hadn't thought of themselves as gamers to some of the fundamental properties that make games fun."

Experts may have believed that games were fundamentally more attractive to males than females, but that difference turns out to have been cultural. Now that smartphones have become game delivery devices, many of the most popular games, such as FarmVille, Kim Kardashian's Hollywood, and Candy Crush, are played by more women than men. All you need is the right environment—and the removal of barriers that prevent novices from taking their first hit—and you'll find a brand-new segment of addicts that looks nothing like the addicts who came before them.

Kimberly Young, a psychologist who practices at a small regional hospital in Bradford, Pennsylvania, coined the phrase "Internet addiction" in 1995, and in 2010 she opened the Center for Internet Addiction—the country's first hospital-based treatment center for Internet addiction. Most Internet addicts are hooked on games. "In the mid-2000s, as the infrastructure of the Internet improved, Internet addiction became a much bigger problem," Young said. "But the biggest changes, by far, were the introduction of the iPhone and then the iPad in 2010." Games became mobile, available to anyone with a smartphone all the time. Instead of a string of teenage boys, Young was suddenly treating both males and females of all ages and personality types. What had saved these people from forming Internet addictions beforehand was that gaming was largely inaccessible. You had to decide to buy a console, and you had to have hours and hours of free time on your hands. Apart from teenage boys, most people were excluded on one or more fronts. "Everybody now has a tablet or an iPhone or a smart device, and it cuts across generations," Young told me. "That's how my career exploded." Young says early lures designed to hook novices are just the beginning. The most compelling experiences maintain their appeal in the longer term, providing benefits not just to beginners but also to veterans.

Miyamoto's Super Mario Bros. appealed to novices, of course, but also contained buried treasure for more experienced players. The game's first level contained a secret tunnel that gave experts a shortcut to the end of the level via an underground chamber filled with coins. The tunnel allowed them to skip Miyamoto's in-game tutorial, and it also rewarded their persistence by playing a string of "ding" sounds as Mario grabbed the underground coins. Because Miyamoto hid some of its charms from all but the game's most devoted fans, many early fans continue to return to Super Mario three decades after its release.

7.

Escalation

According to Google Books there are more than thirty thousand books about "making life easier." These books focus on a wealth of issues, including romantic relationships, managing your finances, succeeding at work, selling on eBay, networking, life as a modern woman, life as a modern man, parenting, losing weight, gaining weight, maintaining your weight, gaining muscle, losing fat, writing exams, making animated films, computer coding, inventing products, getting rich quick, dancing, staying healthy, being happy, living a meaningful life, acquiring good habits, shedding bad habits, and hundreds of other topics. These books suggest that our lives are hard, and that we'd be happier and better off if we could learn to replace hardship with ease. But most of these books weren't written for people enduring major hardships, and there's very little evidence that people with regular lives become happier when you replace challenges with ease.

We know this because people don't seem to embrace ease when you give them a choice.

In the summer of 2014, eight psychologists published a paper in the influential journal *Science* about how people respond when given an opportunity to embrace ease. In one study they asked a group of undergraduate students to sit quietly for ten or twenty minutes. "Your goal," they said, "is to entertain yourself with your thoughts as best you can. That is, your goal should be to have a pleasant experience, as opposed to spending the time focusing on everyday activities or negative things." It's hard to imagine a psychology experiment being less onerous. (The first experiment I ran, almost fifteen years ago, was designed to measure how people behaved when they were sad. I subjected one hundred students to the scene in *The Champ* where a young Ricky Schroder cries as his dad, played by Jon Voight, dies in his arms. The scene is regularly voted the "saddest scene in film," and even the bubbliest students were upset when they left the lab. So asking people to sit quietly with their pleasant thoughts isn't so bad.)

The experimenters added a twist to the experiment. They hooked the students up to a machine that administers electric shocks, and gave them a sample shock to show them that the experience of being shocked isn't pleasant. It isn't agonizing, but it sits somewhere between the pinch of a syringe needle and a bad toothache. Just before leaving the room, the experimenter told the students that the electric shock would be available while they were thinking quietly, that they could experience it again if they wanted to, but that "Whether you do so is completely up to you— it is your choice."

One student—a male, in case you're curious—shocked himself one hundred and ninety times. That's once every six seconds, over and over, for twenty minutes. He was an outlier, but two thirds of all male students and about one in three female students shocked themselves at least once. Many shocked themselves more than once. They'd all experienced the sting of the shock before the experiment, so this wasn't just curiosity. By their own admission in a questionnaire just minutes earlier they didn't find the experience pleasant. So they preferred to endure the unpleasantness of a shock to the experience of sitting quietly with their thoughts. In the experimenters' words, "most people prefer to be doing something rather than nothing, even if that something is negative." As thirty thousand books tell us, we may be looking for an easier life on some level—but many of us prefer to break up a period of mild pleasantness with a dose of moderate hardship.

David Goldhill explained why some degree of hardship is essential. "People don't understand why movie stars are often miserable," Goldhill said. "Imagine getting the girl every night, and never paying for a meal. A game in which you always win, for most people, is boring." The game Goldhill described sounds appealing on its surface, but it gets old fast. To some extent we all need losses and difficulties and challenges, because without them the thrill of success weakens gradually with each new victory. That's why people spend precious chunks of free time doing difficult crosswords and climbing dangerous mountains—because the hardship of the challenge is far more compelling than knowing you're going to succeed. This sense of hardship is an ingredi-

ent in many addictive experiences, including one of the most
addictive simple games of all time: Tetris.

———

In 1984, Alexey Pajitnov was working at a computer lab at the
Russian Academy of Science in Moscow. Many of the lab's scientists worked on side projects, and Pajitnov began working on a
video game. The game borrowed from tennis and a version of
four-piece dominoes called tetrominoes, so Pajitnov combined
those words to form the name Tetris. Pajitnov worked on Tetris
for much longer than he planned because he couldn't stop playing
the game. His friends remember him chain-smoking and pacing
back and forth along the lab's polished concrete floors.

In an interview ten years after the game was released, Pajitnov remembered, "You can't imagine. I couldn't finish the proto-
type! I started to play and never had time to finish the code."
Eventually Pajitnov allowed his friends at the Academy of Science to play the game. "I let other people play, and I realized, it's
not me who's cuckoo! Everyone who touched the game couldn't
stop playing either. People kept playing, playing, playing. My best
friend said, 'I can't live with your Tetris anymore.'" His best
friend, Vladimir Pokhilko, a former psychologist, remembered
taking the game to his lab at the Moscow Medical Institute.
"Everybody stopped working. So I deleted it from every computer. Everyone went back to work, until a new version appeared
in the lab." Pajitnov's boss, Yuri Yevtushenko, who directed the
Computer Center at the Russian Academy of Science, remembered that productivity at the Center plummeted. "The game

was compelling and many of our employees got carried away to the detriment of their work."

Tetris spread from the Academy of Science to the rest of Moscow, and then on to the rest of Russia and Eastern Europe. Two years later, in 1986, the game reached the West, but its big break came in 1991, when Nintendo signed a deal with Pajitnov. Every Game Boy would come with a free game cartridge that contained a redesigned version of Tetris.

That year I saved up and ultimately bought a Game Boy, which is how I came to play Tetris for the first time. It wasn't as glitzy as some of my other favorites, but like Pajitnov I played for hours at a time. Sometimes, as I drifted off to sleep, I'd imagine the blocks tumbling down to form completed rows—a remarkably common experience known as *The Tetris Effect,* which affects people who have played any animated game for long periods of time. Nintendo was smart to include the game with their new portable console, because it was easy to learn and very difficult to abandon. I assumed I'd grow tired of Tetris, but sometimes I still play the game today, more than twenty-five years later. It has longevity because it grows with you. It's easy at first, but as you improve the game gets more difficult. The pieces fall from the top of the screen more quickly, and you have less time to react than you did when you were a novice. This escalation of difficulty is a critical hook that keeps the game engaging long after you've mastered its basic moves. Part of what makes this progression pleasurable is that your brain becomes more efficient as you improve. In fact, in 1991 the *Guinness Book of Records* recognized Tetris as "the first videogame to improve brain functioning and efficiency." That claim was based on research by a psychiatrist named Richard Haier, who worked at the University of California.

In 1991, Haier wondered whether our brains get better at difficult mental tasks with some practice. He decided to watch as people mastered a video game, but he didn't know much about the cutting-edge world of gaming. "In 1991 no one had heard of Tetris," he said in an interview a few years later. "I went to the computer store to see what they had and the guy said, 'Here try this. It's just come in.' Tetris was the perfect game, it was simple to learn, you had to practice to get good and there was a good learning curve."

So Haier bought some copies of Tetris for his lab and watched as his experimental subjects played the game. He did find neurological changes with experience—parts of the brain thickened and brain activity declined, suggesting experts' brains worked more efficiently—but more relevant here, he found that his subjects relished playing the game. They signed up to play for forty-five minutes a day, five days a week, for up to eight weeks. They came for the experiment (and the cash payment that came with participating), but stayed for the game.

One satisfying feature of the game is the sense that you're building something—that your efforts produce a pleasing tower of colored bricks. "You have the chaos coming as random pieces, and your job is to put them in order," Pajitnov said. "But just as you construct the perfect line, it disappears. All that remains is what you fail to complete." Mikhail Kulagin, Pajitnov's friend and a fellow programmer, remembers feeling a strong drive to fix his mistakes. "Tetris is a game with a very strong negative motivation. You never see what you have done very well, and your mistakes are seen on the screen. And you always want to correct them." Pajitnov agreed. "What hits your eyes are your ugly mistakes. And that drives you to fix them all the time." The game allows you the brief thrill of seeing your completed lines flash before they disappear, leaving only your mistakes. So you begin again, and try to complete another line as the game speeds up and your fingers are forced to dance across the controls more quickly.

Pajitnov and Kulagin were driven by this sense of mastery, which turns out to be deeply motivating. In one experiment run by business school professors Michael Norton, Daniel Mochon, and Dan Ariely, students came into a lab and either built a black storage box from IKEA or saw that a pre-built box was already waiting for them. The researchers asked the students how much they'd be willing to pay for the box (with the understanding that they might in fact be asked to pay that amount). Those who had built the box bid a full 63 percent more than did students who happened to be bidding on a box that was pre-built. They were bidding on exactly the same item. This difference—seventy-eight cents versus forty-eight cents—represents the value that people place on their own creations. In another experiment, students were willing to pay five cents for someone else's amateurish origami creation, but twenty-three cents—more than four times as much—for their own (equally amateurish) origami creation. When asked to bid on the origami creations of experts, which were objectively far more impressive, they bid only twenty-seven cents—a mere four-cent premium for a vastly superior product. Other studies have shown that we're also driven to build more Legos when the completed products—the fruits of our Lego-building labor—are stacked up in front of us, rather than removed as soon as they're completed. The sense of creating something that requires labor and effort and expertise is a major force behind addictive acts that might otherwise lose their sheen over time. It also highlights an insidious difference between substance addiction and behavioral addiction: where substance addictions are nakedly destructive, many behavioral addictions are quietly destructive acts wrapped in cloaks of creation. The illusion of progress will sustain you as you achieve high scores or acquire more followers or spend more time at work, and so you'll struggle ever harder to shake the need to continue.

Six decades before Pajitnov released Tetris, a Russian psychologist named Lev Vygotsky was studying how children learn new skills. Like Pajitnov, Vygotsky spent his most productive years in Moscow, at the Moscow State University. He was Jewish, which was a significant handicap for even the brightest aspiring students. But Vygotsky was lucky, winning a spot through the university's annual "Jewish lottery," which decided which Jewish applicants would fill its "no more than 3 percent" Jewish quota. Sadly, Vygotsky would be struck down by a number of illnesses, and he died at age thirty-seven. But he was wildly productive during his brief life, and one of his major contributions explains why Pajitnov and his colleagues were so drawn to Tetris.

Vygotsky explained that children learn best, and are most

motivated, when the material they're learning is *just* beyond the reach of their current abilities. In the classroom context, this means a teacher guides them to clear the hurdle presented by the task, but not so heavy-handedly that they feel their existing skills weren't useful in reaching the task's solution. Vygotsky called this the "zone of proximal development," which he represented with this simple diagram:

THINGS THE LEARNER CAN'T DO AT ALL

ZONE OF PROXIMAL DEVELOPMENT (THINGS THE LEARNER CAN DO WITH HELP)

THINGS THE LEARNER CAN DO WITHOUT HELP

When adults play games, they aren't led along by a teacher— but a well-designed game creates the illusion of being taught. (Remember the first level of Shigeru Miyamoto's Super Mario Bros., which coached novice players through the game's basics.) People who play Tetris, regardless of their abilities, spend most of their time in the zone of proximal development. Like Richard Haier's subjects, they struggle with the game's slowest level until they slowly develop a sense of mastery that allows them to play the second level, and then the third, and so on. The difficulty of the game escalates, but their abilities keep pace—or rather fall just short of mastering the most difficult level they've managed to attain.

The zone of proximal development is deeply motivating. You don't just learn efficiently; you also enjoy the process. In 1990, a Hungarian psychologist named Mihaly Csikszentmihalyi published *Flow*, his classic book on the psychic benefits of mastering a challenge. (One of my professors told me to pronounce Csikszentmihalyi's name as "chick-sent-me-high," which I've always remembered.) Csikszentmihalyi had noticed that many artists became deeply embedded in the business of making art—so deeply that they allowed hours and hours to pass without feeling the need to eat or drink. As Csikszentmihalyi explained, when people experience flow—also known as entering *the zone*—they become so immersed in the task at hand that they lose track of time. Some report a sense of profound joy or rapture when they enter the zone; a rare, long-lasting euphoria that only seems to arise reliably in these rare situations characterized by challenges and the ability to just barely overcome those challenges. (As Csikszentmihalyi acknowledged, flow has been a major part of many Eastern philosophies and religions for centuries. His major contribution was to refine and translate the idea for a new audience.) Csikszentmihalyi created a useful diagram that shows why escalation of difficulty is such a big part of flow:

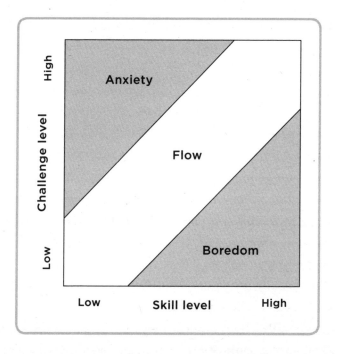

Flow—the channel that runs from the bottom left to the top right of the diagram—describes the experience of tackling a moderate challenge with the skill to master that challenge. Both ingredients are essential. If the challenge is high but you're less skilled, you experience anxiety; if you're skilled but the challenge is low, boredom.

In the context of gaming, experts call this sensation the *ludic loop*—from the Latin *ludere*, for playful. You enter a ludic loop when, each time you enjoy the brief thrill of solving one element of a puzzle, a new and incomplete piece presents itself. The ludic loop can be found in challenging video games, difficult crosswords, repetitive but stimulating work tasks, slot machines that grant you low wins among many losses, and countless other immersive experiences. Ludic loops, like all flow experiences, are very powerful.

When I visited reSTART, the Internet addiction center, I asked one of its founders, Cosette Rae, whether she had ever been addicted to games herself. She was lucky to have been born a few years before the kids she treats today, she said. "Had I been born ten years later I may have developed an addiction. I remember playing a game called Myst. It was beautiful! But it was slow, and it would freeze, and I was, like, I just have too much on my plate." I remember Myst as well. It was a gorgeously rendered role-playing adventure game. It was also very clunky, because the P.C.s of the early 1990s just couldn't handle the demands it placed on their memory chips, and graphics and sound cards. In 2000, a magazine called IGN printed a column titled, "Is the world's best-selling P.C. game ever still worth playing today?" Its conclusion: no. Myst had aged badly, and playing it "was like watching hit TV shows from the 70s. 'People watched that?' you wonder in horror." The patients at reSTART are now playing games that were inspired by Myst and its contemporaries. The big difference is that they're smooth, their graphics are seamless, and they almost never force you to reboot your computer.

What gamers see as progress, Rae sees as dangerous. Her experience with Myst inspired her, fifteen years later, to create artificial barriers that disrupt the formation of ludic loops. She doesn't want to experience flow where games and phones and emails and the Internet are concerned. "When you analyze why people use these gadgets less often, it's when they become irritating—an obstacle. So I used to buy the latest and greatest tech gadgets, the latest and greatest computer software, and I learned, as a harm reduction strategy, to wait two or three years before buying a product. The addict self wants more power and more speed, easier accessibility, the latest and greatest. So I pat my non-addict self on the back and say, 'good job'—you didn't go and buy the new iPhone; you haven't upgraded your computer."

Not everyone avoids temptation so assiduously. Like Alexey Pajitnov thirty years earlier, an Irish game designer named Terry Cavanagh played one of his own games incessantly. Cavanagh is a prolific designer, but he's best known for a game called Super Hexagon. The game belongs to a genre known as "twitch" games, because it requires you to develop almost superhuman reflexes and motor responses. Your task is to guide a small arrow around a circular path at the center of the screen while evading incoming walls for a minimum of sixty seconds. Unlike many compelling games it doesn't coddle you—it's difficult from the very easiest level. (Imagine starting Tetris on Level 8 instead of Level 1.) Even the slowest of the game's six levels is unforgiving, and I had to play for many hours before I completed that level. (I've still never progressed beyond the game's third level.) Super Hexagon is so difficult that some designers call it "masocore"—a game that's almost cruelly —"masochistally"— punishing.

While fine-tuning Super Hexagon in 2011 and 2012, Cavanagh played the game over and over again. He noticed, much as Pajitnov had with early versions of Tetris, that he improved rapidly. What seemed hard initially became easy with practice, and this sense of mastery was addictive. "I think if you can finish the first mode and you're into it, you can finish the game completely," Cavanagh said during an interview. "I've seen this happen with the people that were beta testing it—they thought 'well, this is just way too hard for me' and then they got to the point where their reflexes were good enough and they understood the game well enough that they could actually finish it. That's what the game is all about. It should be a challenge to overcome."

The game was a big hit in the indie gaming community, and it won several major awards in 2012 and 2013. But despite attracting a bevy of fans, Cavanagh had been given a head start, and he seemed to be the best Super Hexagon player in the world. In September 2012, at a conference called Fantastic Arcade, he played the hardest level of the game in front of a large audience. You can watch his astounding performance on YouTube. For seventy-eight seconds he performs a series of agile moves that are hard to see let alone imagine performing yourself. The little arrow jumps around the screen projected behind Cavanagh's head, and the crowd gasps as he conquers the game. He celebrates by saying quietly, quite bashfully, "Now there's a much larger percentage of people who have seen this ending."

At first Super Hexagon sounds too difficult to be appealing, but Cavanagh built in a series of hooks to prevent novices from giving up. The average game lasts only a few seconds at first, and rarely longer than a minute, which means you're never pouring too much time and energy into a single run. "Because it's so short, it's, I hope, kind of inviting," Cavanagh said. "I'm really happy with how that aspect of the game works. You never really feel like you're losing progress, even when you fail at the end of a fifty-nine-second run. You just go right back into it, because the game is tuned in such a way that it doesn't feel like a loss." As soon as the game ends, it starts up again without pause. It doesn't give you time to wallow in defeat, and before you know it you're focusing on a new attempt as though the trail of failures that came before it had never happened. The ludic loop is preserved, and you're never yanked from your flow. The game's music has the same effect. "The music starts in a random place when you restart," Cavanagh said. "If the music started in the very beginning every single time, then every single time you died you'd feel like 'Oh, I've lost and I have to start again from the beginning.' It's really important you don't feel that way, you don't feel like you've lost."

There was something else about Super Hexagon that hooked me: the sense that victory was just around the corner. Sure, my first several hundred attempts ended in failure, but I always felt that, but for a slip of the mouse button, I'd have guided the small arrow away from the oncoming wall. I was sure that I'd finish the level in time. Near wins like these, where you're sure you're close to winning despite falling just short, are very addictive—in fact, often more so than genuine wins.

We know this from a paper that two marketing professors published in 2015. In one experiment they asked a group of shoppers to scratch lottery tickets. Tickets that contained the number eight six times in a row earned the lucky shoppers a twenty-dollar prize. The experimenters designed the tickets so that they either presented a win (left), a near win (center), or a clear loss (right):

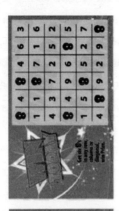

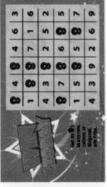

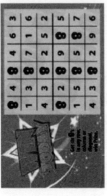

Most shoppers scratched the cards from top left to bottom right, which meant they quickly discovered they had lost in the "clear loss" condition. Shoppers in the other two conditions got off to a great start, but the winners ultimately won and the near winners lost when they reached the critical eighth row. In these and other studies, the experiment's participants completed another activity after either winning or losing the game, while the researchers surreptitiously monitored their behavior. In each case, those who had almost won were more motivated and driven no matter what they were doing. They bought more products from stores, sorted a stack of numbered cards more quickly and efficiently, and walked faster to collect an unrelated reward. The researchers even found that they drooled more—they produced more saliva—following a near win rather than a clear loss. The experience of almost winning lights a fire under us, and drives us to do something—anything—to ease the sense of disappointment that follows a last-minute loss. Other researchers have found similar patterns, suggesting, for example, that gamblers prefer to play games that present near wins on 30 percent of all spins rather than games that present near wins on 15 percent of trials or no trials at all.

Near wins signal that success is nearby. That's why I continued to play Super Hexagon in the face of countless failures. In the context of a game of skill this makes perfect sense—a near win sends the useful signal that you're close to achieving victory. With practice and grit you're likely to achieve that goal. But sometimes that signal is meaningless, particularly when the game relies entirely on luck. As anthropologist Natasha Dow Schüll told me, that's how casinos hook gamblers. Slot machine wins seem to be tantalizingly close, when in fact there's no material difference between a near win and a clear loss. Neither one signals that you're more or less likely to win the jackpot in the future, since it's illegal to change the odds of winning on any particular spin.

The second biggest problem with slot machines is that they lure you in. You can't pass a well-designed slot machine without at least slowing for a quick look. The biggest problem, though, is that they refuse to let you stop playing once you begin. What they do best is to obliterate your stopping rules.

During the 1990s, psychologist Paco Underhill famously watched thousands of hours of retail store security camera footage. The cameras captured all sorts of shopping behavior, most of it mundane but some of it interesting and useful to the store owners who asked Underhill for help. One of Underhill's most famous observations was the so-called *butt-brush effect*. In cluttered stores, where merchandise racks are placed only a few feet apart, customers are forced to squeeze past one another. Underhill's footage captured hundreds of these unintentional butt-brushes, and he noticed an interesting pattern of behavior: as soon as women, and to a lesser extent men, were brushed, they tended to stop browsing and often left the store empty-handed. Butt-brushes were costing stores a lot of money, so he sent a team to investigate why. Were customers abandoning the store as an act of protest? Were they disgusted by the idea of touching a stranger? In fact, customers had absolutely no idea they were reacting to butt-brushes at all. They acknowledged leaving the store, but almost always said it had nothing to do with the presence of other shoppers. Sometimes they cobbled together good reasons for leaving—they were late for a meeting or needed to collect their kids from school—but the pattern was just too strong to deny. What Underhill had identified was a *stopping rule*—a cue that guided customers to stop shopping. The rule wasn't something those customers could explain, but it was there, guiding their behavior all the same.

We tend to overlook stopping rules because often it seems more important in the short-term to question why people start doing something new than why they stop doing something old. If you're selling something, your first question is how you can encourage people to use your product rather than how you can prevent them from moving on to a different one. If you're a doctor trying to encourage your patients to exercise, your first question is how you can get them to start working out—not how you can coax them to continue. And if you're a teacher, your first question is how you can encourage students to study, rather than how you can push them to keep studying over time. You have to ask why people start before you ask why they stop, but stopping rules play a huge and sometimes overlooked role in driving addictive and compulsive behaviors.

Unfortunately, the same new technologies that make life easier also disrupt our stopping rules. Wearable tech like the Apple Watch and Fitbit allow you to track your workouts, but they also discourage you from paying attention to your body's internal exhaustion cues. Both Katherine Schreiber and Leslie Sim, the exercise addiction experts I mentioned earlier, think that wearable tech aggravates the problem. "Tech plays a role inasmuch as it reinforces the calculating mind-set," Schreiber told me. "It reinforces how much attention you pay to walking a certain number of steps or getting a certain number of hours of R.E.M. sleep, for example. I've never used one of these devices because I know they would drive me insane. It's a trigger for all sorts of addictive behaviors." Sim compared Fitbits to calorie counting, which "doesn't help us manage our weight any better; it just makes us more obsessive." Calorie counting makes us less intuitive about what we're eating, and Sim also wondered whether wearable tech made us less intuitive about physical activity. Some of her patients say things like, "if I've only done fourteen thousand steps today, even though I'm really tired and I need to rest, I have to go out

and do my extra two thousand steps." These results are also concerning, because the healthiest approach to exercising in moderation and eating well is to enjoy them—to cultivate an intrinsic
preference for salads and thirty minutes of walking over burgers
and inactivity. Unfortunately, counting calories and steps crowds
out intrinsic motivation by signaling that you're only being
healthy because you're trying to meet numerical targets.

The same technology that drives people to overexercise also
binds them to the workplace twenty-four hours a day. Until recently, people left work behind when they left the office, but now,
with the introduction of smartphones, tablets, remote log-ins, and
emails that find us wherever we happen to be, that stopping rule
is obsolete. Since the late 1960s, but especially in the past two
decades, Japanese workers have whispered about *karoshi*, literally
"death from overworking." The term applies to workers, particularly mid- and high-level executives who struggle to leave work
behind at the end of the day. As a result, they die prematurely
from strokes, heart attacks, and other stress-induced ailments. In
2011, for example, the media described an engineer who died at
his desk at a computer tech company called Nanya. The engineer
had worked between sixteen and nineteen hours per day, sometimes from home, and an autopsy suggested that he died from
"cardiogenic shock."

A recurring theme in karoshi cases is that victims spend far
more time at work than necessary. They're often successful, and
they have more than enough money. They aren't bound to work
longer hours to support themselves, but for one reason or another,
they can't seem to stop. In 2013, Chris Hsee, a business school

professor at the University of Chicago, wrote a paper with three of his colleagues about why people have such weak stopping rules when it comes to work. In one experiment, the researchers gave undergraduate students the opportunity to earn chocolates. During the experiment the students could do one of two things: listen to pleasant, soothing music or endure the harsh sound of an annoying tone. Some of the students earned a chocolate for every twenty times they heard the tone. It was unpleasant and ultimately left the students with chocolates (a sort of wage), so the researchers considered it a form of work. On average the students earned ten chocolates, which seems like a good outcome—until you realize they only ate an average of four chocolates at the end of the experiment. Once they were on the wage-earning treadmill, they couldn't stop even when they had enough in the bank. They were so insensitive to stopping rules that they spent too much time working and not enough time playing. As Kent Berridge, the neuroscientist first introduced in chapter 3, discovered, people sometimes continue wanting a behavior long after it stops bringing them joy. The students, once locked in work mode, couldn't seem to stop even as the benefits of working declined. At the end of their paper, the researchers speculated that:

> Overearning may be an overgeneralized [rule of thumb]. For much of human history, earning rates were low. To earn and accumulate as much as possible was a functional [rule] for survival; individuals did not need to worry about earning too much, because they could not earn too much . . . Like overeating, overearning is a modern-era issue stemming from

advancements in productivity, and it carries potential costs for humans.

You can see the same destruction of stopping rules in other places, too. Until quite recently, gamblers fed dollar bills into slot machines, but now they play with cards that register their wins and losses. Shoppers, similarly, pay for their purchases with credit cards. In both cases, it's hard to keep track of mounting losses that might send a stopping signal if they were more obvious. Instead of watching as the wads of bills in their wallet dwindle, shoppers and gamblers use a single card that remotely and abstractly registers each loss and each expense.

In a classic paper, marketing professors Dražen Prelec and Duncan Simester showed that people will pay up to twice as much for the same item when using a credit card rather than cash. Credit cards, like slot machine cards, hide all feedback from a spender, who has to keep track of his own gains and losses instead. American Express once coined the slogan, "Don't leave home without it," but Prelec and Simester cleverly turned that slogan on its head when they titled their paper "Always Leave Home Without It."

I heard a similar story time and again from game designers, who described a growing movement of ethical game design. NYU Game Center director Frank Lantz told me that FarmVille and other Facebook games were successful in part because once you were hooked they never let you go. "Facebook games run twenty-four hours a day—they're persistent games. They aren't games where you have to start a session, and then play, and then save your results, and then come back later and begin the session again. They're just always going whenever you want to play them." The fun never ends because the game doesn't impose its own stopping rule. There are no chapters or sessions or levels that tell you when your gaming session begins and when it ends. Bennett Foddy agreed: "Some designers are very much against infinite format games, like Tetris, for example, because they're an abuse of a weakness in people's motivational structures—that they won't be able to stop. Instead they prefer to make games that engage you till you get to the end—and then it's over and you're free from it."

Some games pay vague lip service to this idea by warning you to stop and take a break when you've been playing for a long time. But those warnings are toothless and, in some sense, tease you to keep playing. I played a strategy game called 2048, which was all the rage on New York City subways for a couple of years. (I discovered the game by asking a fellow subway rider—the tenth person I'd seen playing the game in a matter of days— what he was playing.) One of the game's welcome screens says "Thank You. Enjoy the game and don't forget to take a break, if needed!" Right below that warning is a button that takes you to Apple's app store, where you're offered a medley of similarly addictive games, many by the same design team. The solution, as far as the designers of 2048 are concerned, is to stop you from playing one game by offering you a series of others to take its place.

As with Tetris and 2048, humans find the sweet spot sandwiched between "too easy" and "too difficult" irresistible. It's the

land of just-challenging-enough computer games, financial targets, work ambitions, social media objectives, and fitness goals.
Addictive experiences live in this sweet spot, where stopping rules
crumble before obsessive goal-setting. Tech mavens, game developers, and product designers tweak their wares to ensure their
complexity escalates as users gain insight and competence.

8.

Cliffhangers

A minibus veers off a mountain road and teeters on the edge of a cliff. The minibus is an empty shell without seating. Inside are eleven thieves and their pile of stolen gold. The men hug the back wall as the gold slowly slides away from them, tipping the minibus toward oblivion. One of the men crawls slowly toward the gold. The only sounds are his shuffling, the creaking minibus, and the whistling of alpine winds. He moves within two feet of the gold, but the bus tips farther forward and it slides beyond his reach. Then, he rolls onto his back, faces his companions, and says calmly: "Hang on a minute, lads. I've got a great idea." The story ends.

In the summer of 1969, thousands of cinemagoers enjoyed the first ninety-four minutes of *The Italian Job*, but many hated this, the final ninety-fifth minute. In their own words, the ending was "ridiculous," "pretentious garbage," "horrible," "crap," "frustrating," "not funny," "without morals," "without heart," "a turkey," "like a soft drink that's gone flat," "enjoyable maybe if you've had a lobotomy." It takes a special ending to inspire this sort of vitriol, and that ending turns out to have been no ending at all: a literal and metaphorical cliffhanger. The problem here was that viewers had committed an hour and a half to the story, and like all humans they were wired for closure. If you've ever been denied a joke's punchline, you'll know that it's better to hear no story at all than to hear all but the story's final beat.

F orty years earlier, a Lithuanian psychologist named Bluma Zeigarnik stumbled on the power of cliffhangers. She was sipping coffee at a small café in Vienna when she noticed that her waiter remembered his customers' orders with superhuman clarity. When he approached the kitchen, he knew to tell the cook to prepare eggs Benedict for table seven, a ham and cheese omelet for table twelve, and scrambled eggs for table fifteen. But as soon as those orders landed at tables seven, twelve, and fifteen, his memories of them vanished. Each order presented the waiter with a miniature cliffhanger that was resolved when the right meal reached the right customer. Zeigarnik's waiter remembered his open orders because they wouldn't leave him in peace—they nagged at him in the same way that the teetering bus nagged at *The Italian Job*'s frustrated viewers. When the waiter served each order, the cliffhangers were resolved, and his mind was free to focus on the new cliffhanger presented by his next order.

Zeigarnik designed an experiment to uncover the effect more

carefully, inviting a group of adults into her lab to work on twenty different brief tasks. Some of these were manual, like creating clay figurines and building boxes, and others were mental, like arithmetic sums and puzzles. Zeigarnik allowed her participants to complete some of the tasks, but she interrupted them before they could complete others, and forced them to move on to the next task. The subjects were loath to stop, and they sometimes objected quite strenuously. Some were even angry, which showed how much tension Zeigarnik introduced with her interruptions. At the end of the experiment, she asked them to remember as many of the tasks as they could.

The results were striking. Like the waiter in Vienna, her participants recalled about twice as many unfinished tasks as they did finished ones. At first, Zeigarnik wondered whether the unfinished tasks were more memorable because participants experienced a small "shock" when they were interrupted. But when she ran a similar experiment, again interrupting her participants as they completed some tasks but then allowing them to complete those tasks later, the effect vanished. It wasn't interruption that made the tasks memorable, but rather the tension from not being able to complete them. In fact, the interrupted tasks that were later completed were no more memorable than the tasks completed without interruption. Zeigarnik summarized her results: "When the subject sets out to perform the operations required by one of these tasks there develops within him a quasi-need for completion of that task. This is like the occurrence of a tension system which tends towards resolution. Completing the task means resolving the tension system, or discharging the quasi-need. If a task is not completed, a state of tension remains and the quasi-need is unstilled." So the Zeigarnik Effect was born: incomplete experiences occupy our minds far more than completed ones.

Once you look for it, the Zeigarnik Effect is everywhere. Take the case of earworms—catchy songs that stubbornly play over and over inside your head. Jeff Peretz, a guitarist and music professor at New York University, told me that some earworms achieve cult status because they plant cliffhangers that never resolve. He pointed to the colossal 1978 hit song "September," by Earth, Wind & Fire, a combination of percussive bounce and brassy punch that begins with the line, "Do you remember the twenty-first night of September?" In 2014, as the song turned thirty-six, longtime band member Verdine White told an interviewer that, "People now are getting married on September 21st. The stock market goes up on September 21st. Every kid I know now that is in their 20s, they always thank me because they were born on September 21st. They say it's one of the most popular songs in music history right now."

This was the golden age of disco, and in many ways, "September" is a model disco classic. But in other ways it's very unusual. Many pop hits follow a standard circular chord progression—they launch like a rocket ship, hover for a time above the launch pad, and ultimately close the melodic loop by returning to Earth. In the world of Bluma Zeigarnik's waiter, these tracks are fulfilled orders: they're satisfying, but your mind leaves them behind when they end, and another song begins.

Not so for "September," according to Peretz. "One of the amazing things about the chord progression in 'September' is that it never lands. It makes this loop that you never want to stop hearing. And that's why it's so popular still, to this day. This same approach is used for the song's main theme, its chorus, and its hook. It keeps going on and on. Without a doubt this contributes to its longevity. It has all the makings of an earworm. And this looped feature only makes it harder to leave once it does get stuck in your head." Long after we've forgotten other songs, the endless loop continues to demand our attention. Almost forty years after its release, "September" remains a staple at parties and weddings. (By coincidence, my wife and I were married on the evening of September 21, 2013, and our D.J. was under strict instruction to include the song in his playlist.)

"September"'s cliffhangers never quite resolve, but some songs endure in our minds because they resolve their cliffhangers in unexpected ways. In the summer of 1997, Radiohead released their cult track "Karma Police," which showcased the band's musical sophistication. The song uses two subtly different versions of the same melody, and until you've listened to it many times, you have no idea which version you're about to hear. There's no rhyme or reason that guides you, and so, Peretz explains, it keeps you on your toes. "The song has you wondering which version of the loop you're going to hear. It seems too sophisticated to be an accident, and I imagine when [lead singer] Thom Yorke was writing the song, he had in mind the idea of karma being a cyclical thing. He totally rung the bell with that one. It's an iconic song. Stevie Wonder's song "Evil" is similar. It has a sequence that starts out in C major, but when it brings you back around to where you started, you're in a new place. It doesn't bring you home."

"September" runs for a gripping three minutes and thirty-five seconds, but it pales next to a category of addictive experiences that grip audiences for months at a time.

In October 2014, National Public Radio began broadcasting *Serial,* a twelve-part podcast that ran for two and a half months. A team of journalists led by NPR's Sarah Koenig were investigating whether a Baltimore high school student named Adnan Syed had been wrongly convicted for his ex-girlfriend Hae Min Lee's murder in 1999. Other podcasts had developed a following, but *Serial* was wildly and uniquely popular. (When I emailed Koenig for an interview, she very politely declined my request. "I'm afraid I just can't," she told me. "I'm sort of deluged at the moment.") For three months, countless conversations included the question, "Have you heard about *Serial*?" I discussed the podcast with friends and strangers everywhere, and I wasn't alone. A number of major publications wrote about *Serial*'s success, and many of their titles and opening paragraphs focused on the podcast's addictiveness:

> The host of this compelling, addictive nonfiction murder-mystery talks about the show's origins and why it's okay to "like" her interviewees.
>
> —*Rolling Stone*

> The thirteen stages of being addicted to "Serial."
>
> —*Entertainment Tonight*

"Serial": The Highly Addictive Spinoff Podcast of "This American Life."

—NBC News Online

Ira Glass and the folks behind *This American Life* radio recently launched "Serial," an addictive podcast about a gruesome murder and the curious court case that convicted a 17-year-old kid. And it's better than the best episode of *Law & Order* because it features the actual people who lived through the tragedy—plus, you have no clue how it's going to end.

—Entertainment Weekly

This last quote nails *Serial*'s magic ingredient: Koenig and her team opened a Zeigarnik loop, but none of her listeners knew when (if ever) the loop would be closed. Would the true murderer be revealed in episode three? Episode nine? In the final episode? Never? Halfway through the series, Koenig admitted that she had no idea how the podcast would end. After a year of interviews and careful research, she and her team were no closer to resolving the only question that really mattered: who killed Hae Min Lee? The audience was rapt because the answer always seemed within reach. Many episodes included one or two interviews with Syed, the convicted murderer. He always seemed to be on the verge of saying something incriminating— or of saying something that would prove his innocence beyond doubt. And the same was true of countless other interviews. One of Syed's acquaintances provided an alibi that seemed to place him at a library precisely when the murder was supposed to be occurring several miles away. But that lead broke down, and the loop remained open.

Thousands of listeners downloaded the final podcast on December 18, 2014, hoping for an answer. But none came. Koenig believed Syed was innocent, but she admitted she wasn't completely sure. The show ended, but the cliffhanger remained, and the listeners refused to move on. They established vibrant online discussion groups. The guilty camp scolded the innocents for being naïve, and the innocents called the guilties jaded skeptics. Almost fifty thousand *Serial* fans shared their views on a *Serial* page (or subreddit) established on the Reddit website. The best evidence that their engagement rose above simple interest came on January 13, 2015. This was the sixteen-year anniversary of Hae Min Lee's murder, and the subreddit's moderators honored Lee by suspending the site for twenty-four hours. In its place was a short message:

On January 13, 1999 life would be changed forever.

Hae Min Lee was an extraordinary individual.

. . .

It was 16 years ago today that her life was ended tragically, and her family and friends' lives would never be the same.

While Hae's murder was the basis of the podcast *Serial*, let us never forget the tragedy itself.

Out of respect for Hae's memory, this subreddit will fall silent for a day so that we can all reflect on the true injustice at the center of a heated debate.

Many users applauded the tribute, but others went into *Serial* withdrawal. A user named hanatheko admitted, "Wow, I am addicted . . . the past 24 hours were painful and I fell ill with depression." For hanatheko, a day without the site was a day too long. Others felt the site's moderators had no right to shutter the site for any reason at all. One user suggested these angry users were "acting like the Westboro Baptist Church of the fucking Internet right now." Another named Muzorra pointed out that "all the commentary . . . that the victim always gets lost and becomes a point of data and little more . . . gets forgotten the moment someone makes it a little harder to get at their toy for a little while." When the site went live again at midnight, hanatheko, Muzorra, and thousands of other users went back to attacking and defending Camp Guilty, Camp Innocent, and Camp Undecided.

NPR's release of *Serial* heralded a flood of unsolved real-life crime documentaries. In February 2015, HBO released *The Jinx*, which tracks the life of Robert Durst, a man who was associated with a number of unsolved murders. The day before HBO released the documentary, Durst was arrested for one of those murders—fueled in part by some of writer Andrew Jarecki's discoveries. Then, in December 2015, Netflix released a ten-part real-life murder documentary called *Making a Murderer*. The documentary's filmmakers, Laura Ricciardi and Moira Demos, spent ten years tracking a man named Steven Avery, who had been convicted of murdering a young woman in small-town Wisconsin. *The Jinx* and *Making a Murderer* were just as addictive as *Serial*, and both attracted waves of acclaim and millions of viewers. All three programs are produced with skill—but much of their popularity is baked into their ambiguity. In *Slate*, Ruth Graham wrote about *Making a Murderer*:

> "This is the perfect *Dateline* story," a *Dateline* producer says of the Avery case in *Making a Murderer*. "It's a story with a twist, it grabs people's attention. . . . Right now, murder is hot." . . . But if *Dateline* leaves viewers hanging over commercial breaks, the multiepisode format of shows like *Making a Murderer* dangles us over much deeper chasms. The series may be more prestige than pulp, but it delivers the same pleasure-pellets of any crime story: "That poor woman!" "Who really did it?" "Someone must pay!"
>
> Take Episode 4 of *Making a Murderer*, which ends with a whopper of a plot bombshell . . . Cue my husband and I freaking out on the couch and agonizing over whether to stay up late to watch another episode. With a cliffhanger like that, how could we not?

As I write this, people are still feverish about *Serial* and *Making a Murderer*. (*The Jinx* has a devoted following, too, though it's perhaps tempered in part by Durst's arrest and the documentary's more limited release.) The *Serial* and *Making a Murderer* subreddits continue to attract new posts each day. But if someone can prove Steven Avery's innocence, or who murdered Hae Min Lee, the loops will close, and the subreddits will wither. A cliffhanger only lasts until you know whether the bus plunges, a waiter only remembers an order until the plate reaches his customer, and the fate of a mobster from suburban New Jersey remains interesting only as long as you don't know whether he lives or dies.

When David Chase wrote the eighty-sixth and final episode of *The Sopranos*, he posed a question that he refused to answer: was Tony Soprano dead? For eight years New Jersey mob boss Tony Soprano evaded death while ninety-two of his enemies and friends faded away. They died from gunshot wounds and beatings and drowning and natural causes; from stabbings and heart attacks and strangulation and drug overdoses. Their deaths captivated viewers, but nowhere near as much as Tony's purgatory absorbed them.

The scene is legend. On June 10, 2007, twelve million Americans watched as Tony Soprano and his family gathered at Holsten's diner. A man in a brown leather jacket enters the restaurant, and sits at the counter. He glances back at the family, briefly, and heads for the restroom. In the show's final seconds, a bell on the front door dings, Tony looks up toward the door, and the screen cuts to black. For eleven seconds it remains that way, eight years of action reduced to a profound quiet. Many viewers wondered whether their TVs or cable boxes had cut out at exactly the wrong moment, but this was Chase's vision.

Fans of the show were perplexed, so they took to Google. The search engine hosted a flood of searches for "Sopranos final episode" beginning at 10:02 P.M. on the East Coast, which continued well into the night. In their desperate search for some kind of resolution, viewers hoped someone out there on the web was more sophisticated than they were. (Eight years later, *Serial* fans would do the same when they took to Reddit.) Media critics either loved the episode or hated it, and without fail saved most of their energy for its closing five minutes. What had happened? Why had Chase cut the story short?

Two competing theories surfaced. On the one hand, perhaps Chase was trying to suggest that life for Tony and his family would continue beyond the show's end. Early in the final scene, Tony had popped a couple of coins in a small jukebox at his table, and Journey's "Don't Stop Believin'" began to play. The last thing viewers heard was singer Steve Perry launching into the song's chorus, "Don't stop . . . !" Chase refused to let Perry complete the phrase, and perhaps the two words that closed the show served as a message: the show had ended, but the lives it depicted wouldn't stop.

On the other hand, many fans were convinced that the silent black screen signaled Tony's death. Since Tony wasn't alive to experience the world after his death, viewers were treated to the same abrupt end. His wife and kids would live to hear Steve Perry sing the final word in the song's title, but it might be drowned out by the gunshot that ended Tony's life. According to this theory, the man in the leather jacket was Tony's assassin; in an homage to Tony's favorite scene from *The Godfather*, perhaps the man had gone to the bathroom to retrieve a gun. If Chase were implying that Tony was dead, he couldn't have chosen a more apt final word than "stop!"

TV journalists clamored for an answer, and Chase occasionally tossed a crumb or two in reply. He continues to lead them on, and refuses to offer a definitive interpretation. In his first interview after the show ended, he said, "I have no interest in explaining, defending, reinterpreting, or adding to what is there. No one was trying to be audacious, honest to God. We did what we thought we had to do. No one was trying to blow people's minds, or thinking, 'Wow, this'll piss them off.'" Eight years and several interviews later, fans were still unsatisfied. In April 2015 Chase told a writer that, "It was very simple and much more on the nose than people think. Either it ends here for Tony or some other time. But in spite of that, it's really worth it. So don't stop believing." In some interviews, he seemed confused by the question. "I saw some items in the press that said, 'This was a huge fuck you to the audience.' That we were shitting in the audience's face. Why would we want to do that? Why would we entertain people for eight years only to give them the finger?"

Serial fans were more disappointed than angry, because Sarah Koenig wanted to know who killed Hae Min Lee as badly as they did. She was on their team. But Chase was an antagonist, willfully denying his viewers an answer to the most important question he'd posed in eight years. The *Chicago Tribune*'s Maureen Ryan spearheaded the "pissed off" camp in her column titled, "Are you kidding me? That was the ending of 'The Sopranos'?" She told her readers, "You can call the ending sadistic. You can call it an ending that leaves room for a sequel. Either way, it'll have fans talking for months." One commenter named Ryan agreed. "The finale sucked! The final shot ruined the entire episode for me. We were robbed . . . ROBBED, I tell you!" And yet, for all their anger, nearly a decade on people can't stop talking about the show's final episode. It's as though they've taken the show's final two words from Steve Perry too seriously: "Don't stop!"

Which of the following steps in the chain below would you expect to make people happiest?

> Step 1: Desiring something (food, sleep, sex, etc.).
> Step 2: Wondering whether that desire will be satisfied.
> Step 3: Having the desire satisfied.
> . . . Repeat for the next desire.

Step 3 is the obvious answer. It's the step that frustrated fans when *The Italian Job*, *Serial*, and *The Sopranos* ended without resolution, and it's the reason we bother with steps 1 and 2 at all. But, in 2001, Greg Berns and three neuroscientist colleagues undertook a study that asked twenty-five adults to put a small tube in their mouths as they lay on their backs in an fMRI machine. The machine scanned their brains for evidence of pleasure as an experimenter fed them drops of water and fruit juice through the tube. Most of the adults preferred juice to water, but the human brain treats both juice and water as small rewards. For half of the experiment, the drops came in predictably, every ten seconds, alternating between water and juice:

Then, during the other half, the experimenters introduced the element of surprise. Now the adults had no idea when they'd receive their next reward, or whether it would be juice or water:

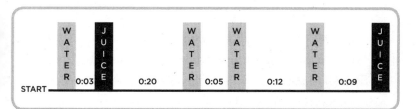

If satisfaction were all that matters, participants' brains would have fired identically in both halves of the experiment—or perhaps more vigorously in the predictable half, when they could anticipate and savor the coming reward. But that's not what happened. Predictability is pleasing at first, but it loses its luster. Near the end of the predictable half of the experiment, participants' brains began responding more and more weakly.

Not so during the unpredictable run, which hooked participants in the same way that *Serial* hooked its listeners. When the rewards were unpredictable, participants enjoyed them that much more—and continued to enjoy them through to the end of the experiment. Each new reward followed its own micro-cliffhanger, and the thrill of waiting made the entire experience more pleasurable for a longer period of time.

These same micro-cliffhangers drive the thrill of compulsive shopping. In 2007, a team of entrepreneurs introduced a remarkably addictive online shopping experience called Gilt. Gilt's website and app promote flash sales that last between one and two days each. Sales are available only to members, and they feature well-priced designer clothes and home goods. The platform is booming, with six million members, so its merchandisers can purchase huge quantities of heavily discounted high-end products. Even after the site tacks on a small margin per item, members pay far less than retail prices.

New sales arrive without warning, so members constantly refresh their pages. Each newly loaded page produces a micro-cliffhanger. For many of Gilt's members, the site offers a low-grade thrill amid their otherwise predictable lives. You can see this in the spike of lunchtime traffic between noon and one every afternoon, during which the site sometimes draws in more than a million dollars in revenue.

Darleen Meier, who runs a lifestyle blog called *Darling Darleen*, was excited when her membership was approved in October 2010. (She was on a wait list for several weeks beforehand.) Meier treated her readers to a front-row seat, celebrating her membership and then sharing some of her favorite purchases. But, just two months later, Meier was moved to publish a post titled "Gilt Addict." The problem became clear when she barely resisted buying a well-priced Vespa scooter. (She suppressed the impulse after imagining how her husband would respond when he saw the scooter.) Meier's relationship with Gilt intensified when a chime began alerting her when a new deal had landed on the site. Regardless of what she was doing, she'd stop to check the app. Sometimes, she found herself pulling off the road while running an errand or driving to pick up her young son from school. Sometimes the cliffhanger didn't resolve in Meier's favor—some of the deals didn't appeal to her—but often, by the time the car was moving again, she'd spent hundreds or even thousands of dollars.

At the height of her Gilt addiction, new boxes were landing on her doorstep every day.

Meier wasn't alone. Online message boards were full of shopping addicts searching for help. On PurseForum, a social network for avid shoppers, Cassandra22007 admitted to being addicted to Gilt, and to other so-called flash sale websites:

> It's become painfully clear to me that I have a problem with Gilt Group and I need an intervention! I'm thinking about banning myself from this site at least temporarily. Basically, I'm unemployed right now and I really have no excuse for buying new clothes and stuff that I probably won't wear until I'm employed again. I currently have like 6-10 items I've gotten there that I have not actually worn/used yet, and I just ordered like 5 more things today.

What's striking about Cassandra22007's behavior is that she wasn't buying clothes because she needed them. Just as Greg Berns had shown with his juice experiment, it wasn't so much the reward itself that mattered, but rather the thrill of the chase. Gilt didn't provide shoppers like Meier and Cassandra22007 with products they couldn't get elsewhere—it provided them with a string of micro-cliffhangers that made the act of hunting down those products deeply addictive.

This shopping produces a lot of clutter, and there's now a cottage industry of self-styled home organization gurus. The latest is Marie Kondo, a Japanese "cleaning consultant." Kondo practices a method that she calls KonMari: throwing out everything in your home that doesn't "spark joy." Kondo explained the principles of KonMari in *The Life-Changing Magic of Tidying Up*, which she first published in 2011. The book has been translated into dozens of languages, and has sold more than two million copies worldwide. Kondo has since published a companion volume, *Spark Joy*, which is also a major bestseller. Tidying up isn't easy, because it goes against the human instinct to retain value. We hate throwing something out if it might provide future value, and it's hard to know for sure that once-useful possessions won't be useful again. But KonMari has one tremendous asset: tidying up is a sort of open loop that demands closing. We hate to throw things out, but we also hate clutter. The people who shop obsessively become the same people who tidy obsessively, and the process becomes a self-perpetuating loop. Once you know to look, you start seeing loops like this one everywhere.

In August 2012, Netflix introduced a subtle new feature called "post-play." With post-play, a thirteen-episode season of *Breaking Bad* became a single, thirteen-hour film. As one episode ended, the Netflix player automatically loaded the next one, which began playing five seconds later. If the previous episode left you with a cliffhanger, all you had to do was sit still as the next episode began and the cliffhanger resolved itself. Before August 2012 you had to decide to watch the next episode; now you had to decide to *not* watch the next episode.

At first this sounds like a trivial change, but the difference turns out to be enormous. The best evidence of this difference comes from a famous study on organ donation rates. When young adults begin driving, they're asked to decide whether to become organ donors. Psychologists Eric Johnson and Dan Goldstein noticed that organ donations rates in Europe varied dramatically from country to country. Even countries with overlapping cultures differed. In Denmark the donation rate was 4 percent; in Sweden it was 86 percent. In Germany the rate was 12 percent; in Austria it was nearly 100 percent. In the Netherlands, 28 percent were donors, while in Belgium the rate was 98 percent. Not even a huge educational campaign in the Netherlands managed to raise the donation rate. So if culture and education weren't responsible, why were some countries more willing to donate than others?

The answer had everything to do with a simple tweak in wording. Some countries asked drivers to opt in by checking a box:

If you are willing to donate your organs, please check this box: ☐

Checking a box doesn't seem like a major hurdle, but even small hurdles loom large when people are trying to decide how their organs should be used when they die. That's not the sort of question we know how to answer without help, so many of us take the path of least resistance by not checking the box, and moving on with our lives. That's exactly how countries like Denmark, Germany, and the Netherlands asked the question—and they all had very low donation rates.

Countries like Sweden, Austria, and Belgium have for many years asked young drivers to opt out of donating their organs by checking a box:

If you are NOT willing to donate your organs, please check this box: ☐

The only difference here is that people are donors by default. They have to actively check a box to remove themselves from the donor list. It's still a big decision, and people still routinely prefer not to check the box. But this explains why some countries enjoy donation rates of 99 percent, while others lag far behind with donation rates of just 4 percent. After August 2012, Netflix viewers had to opt out of watching another episode. Many chose to do nothing and, slack-jawed, they began their eighth consecutive episode of *Breaking Bad*.

Netflix subscribers had been binge-watching since the company introduced streaming in 2008, but bingeing has been escalating since then. Google Trends, which measures the frequency of Google searches across time, shows the frequency of searches for "binge-watching" between January 2013 (when people first begin searching for a term) and April 2015 in the United States:

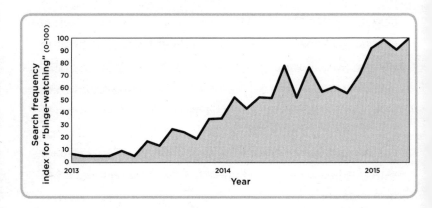

And this one shows the frequency of searches for "Netflix binge" for the same time period in the United States:

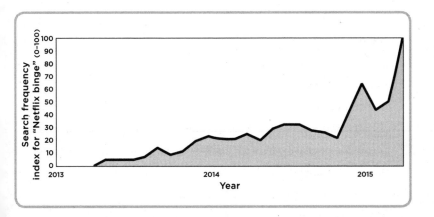

Search term popularity is an indirect measure, but Netflix conducted its own research in November 2013. The company employed a market research firm to interview over three thousand American adults. Sixty-one percent of these people reported some degree of binge-watching, which most respondents defined as "watching between two and six episodes of a TV show in one sitting." Netflix found similar patterns in viewing data, which it collected from 190 countries between October 2015 and May 2016. Most people who binge complete the first season of the shows they're watching in four to six days. A season once stretched on for months at a time, but now it's consumed in under a week, at an average of two to two and a half hours a day. Some viewers report that binge-watching improves the viewing experience, but many others believe that Netflix—and post-play in particular—

has made it very difficult to stop watching just one episode at a time. Much of this rise, charted in the Google Trends graphs, reflects the effectiveness of cliffhangers, and the absence of barriers between the end of one episode and the beginning of the next.

When Willa Paskin, *Slate*'s television critic, reviewed a show called *Love*, she explained that even mediocre TV shows become addictive with "an assist" from binge-viewing. *Love* was a Netflix production, released in a single batch of ten episodes:

> The show gets an assist from binge-watching itself—a style of viewing that encourages audiences to invest in the characters as people, regardless of how little artistry surrounds them. It's like being told a story, any story: At a certain point, you just want to know what happens next. If *Love* aired every week, you could take it or leave it. But Netflix makes it so easy to watch three episodes in one sitting that it's tempting to keep plowing forward on the force of curiosity alone—just how are these crazy kids going to get together? Binge-watching provides a show without much plot all the necessary momentum. By the time you stop hurtling forward, you've already seen it all.

Bluma Zeigarnik, the psychologist we met earlier in this chapter, lived a long and remarkable life littered with cliffhangers. In 1940, her husband Albert was sentenced to ten years in a Soviet prison camp on the charge of spying for Germany. Zeigarnik was left to wonder where he was and when he might come home. When the Soviet authorities captured Albert, they left behind a document that explains why we know so little about Zeigarnik's life. That document, which her grandson stumbled on many decades later, states that the authorities had seized "the contents of a sealed room with numerous documents, folders, notebooks, and records."

Zeigarnik's career took off eventually, but her academic life was just as turbulent as her personal life. She was forced to write three doctoral dissertations after the Soviet authorities refused to recognize her first dissertation, and her second was stolen. She had copies of the second dissertation, but was forced to destroy them when she feared that the thief might publish her work and accuse her of plagiarism. For almost thirty years, Zeigarnik wandered in academic purgatory before completing her third dissertation and joining Moscow State University as a psychology professor in 1965. She was elected chair of the department two years later, and held that position for the next two decades, until her death. With mountainous talent and dogged determination, Zeigarnik ensured that the cliffhanger ultimately resolved in her favor.

9.

Social Interaction

In December 2009, best friends Lucas Buick and Ryan Dorshorst began selling an iPhone app. The app sold for $1.99, and the pair watched eagerly as the download counter climbed. Thirty-six hours after its launch it was the most downloaded app in Japan. Sales rose more slowly in the U.S. but by New Year's Day U.S. customers had downloaded over 150,000 copies of the app. Apple itself took notice, and soon the app was front-and-center on the Apple Store's homepage.

The app was called Hipstamatic, and it allowed iPhone users to digitally manipulate the photos they took with the camera built into their phones. With the help of digital film, flashes, and lenses, even naïve photographers could turn mundane shots into masterpieces that mimicked the retro snaps of the 1980s. Experts were paying attention, too. Damon Winter, a *New York Times* photographer, used the app to take shots of soldiers in Afghani-

stan in 2010. The photos won Winter third place in the Pictures
of the Year International photojournalism competition, and further enriched the Hipstamatic brand.

Buick and Dorshorst were graphic designers by trade, but
they also happened to be intuitive entrepreneurs. To cultivate the
app's retro appeal, they used names like *Ina's 1982 Film*, *Roboto
Glitter Lens*, and *Dreampop Flash*. Their masterstroke was inventing a rich backstory for the app that journalists have since
struggled to authenticate. As they told it, in 1982 two brothers
from Wisconsin created a camera named the *Hipstamatic 100*.
Their aim was to build a camera that was cheaper than its film,
and though they succeeded, they managed to sell only 154 units.
The brothers died in a tragic car accident in 1984, and their older
brother, Richard Dorbowski, kept the three remaining Hipstamatic 100s in his garage until July 29, 2009, when Buick and
Dorshorst told him they wanted to release a digital version of the
camera.

Journalists were captivated by the story, and they described
Hipstamatic's romantic history in dozens of feature stories. They
were helped by scattered online evidence to support the story: a
blog page on the Hipstamatic 100 written by Dorbowski (with
photos of his younger brothers in the early 1980s), and Facebook
and LinkedIn pages that described Dorbowski as living in Wisconsin and working as chief comptroller at a paper company. It
wasn't until several years later, when other journalists tried to
delve deeper, that the backstory crumbled. The three brothers
were a figment, and so, apparently, was the Hipstamatic 100.
Still, the Hipstamatic app was real, and hundreds of thousands of
copies were selling every month. Apple crowned Hipstamatic

"2010's App of the Year," and the *New York Times* included the app in its "Top Ten Must-Have Apps for the iPhone" list in November 2010.

Buick and Dorshorst were riding high, but a pair of entrepreneurs living in San Francisco were preparing to release a rival app. Kevin Systrom and Mike Krieger launched Instagram in October 2010. The two apps offered the same basic service, so entering the market ten months late put Instagram at a huge disadvantage. Though Instagram lacked Hipstamatic's charming backstory—its name simply combined the words "instant" and "telegram"—Systrom and Krieger were canny businessmen. If 2010 was the year of Hipstamatic, 2011 was the year of Instagram. Hipstamatic remained popular, but its download count slowed, and Instagram soon had a larger base of users. Having crowned Hipstamatic "App of the Year" in 2010, Apple bestowed the same honor on Instagram in 2011. In 2012, Hipstamatic's user count peaked around five million, and Instagram now has around three hundred million users. But the biggest difference between the apps came on April 9, 2012, when Facebook acquired Instagram for one billion dollars. When Dorshorst read about the acquisition, he was convinced he was reading a headline from satirical newspaper *The Onion*. He had to double-check. Laura Polkus, a former designer at Hipstamatic, remembered, "We saw Mark [Zuckerberg]'s blog post, and it was like, "Wait, one billion? Like, a billion dollars? What does that mean for us? Does that mean Instagram won?"

Hipstamatic and Instagram offered the same core features, so why did Hipstamatic falter while Instagram continues to grow? The answer lies in two critical decisions that Systrom and Krieger made before they released the app. The first was to make the app free to download. That got users in the door, and it explains in part why so many users downloaded the app early on: there was no risk of spending on a dud, so at worst they could delete the app a couple of days later. But many apps are free, and they still fail miserably. It was the pair's second decision that made the difference: Instagram users posted their photos on a dedicated social network tied to the app. (Hipstamatic users could upload their photos on Facebook, for example, but Hipstamatic wasn't itself a stand-alone social network.)

It's easy to see why Zuckerberg chose to acquire Instagram. He and Systrom shared a similar insight: that people are endlessly driven to compare themselves to others. We take photos to capture memories that we'll revisit privately, but primarily to share those memories with others. In the 1980s, that meant inviting friends over to watch slides of your recent vacation, but today that means uploading photos of your vacation in real time. What makes Facebook and Instagram so addictive is that every activity you post either does—or doesn't—attract *likes*, *regrams*, and comments. If one photo turns out to be a dud, there's always next time. It's endlessly renewable because it's as unpredictable as people's lives are themselves.

So what is it about Instagram's social feedback mechanism that makes it so addictive?

———

People are never really sure of their own self-worth, which can't be measured like weight, or height, or income. Some people obsess over social feedback more than others do, but we're social beings who can't ever completely ignore what other people think of us. And more than anything, inconsistent feedback drives us nuts.

Instagram is a font of inconsistent feedback. One of your photos might attract a hundred likes and twenty positive comments, while another posted ten minutes later attracts thirty likes and no comments at all. People clearly value one photo more than the other, but what does that mean? Are you "worth" a hundred likes, thirty likes, or a different number altogether? Social psychologists have shown that we adopt positive ideas about ourselves more readily than we adopt negative ideas. To see how this works, answer the following questions quickly, without giving them too much thought:

Below you'll see a list of personality traits. Please estimate the percentage of people in your town who embody *less* of each trait than you do:			
sensitive	sophisticated	ingenious	disciplined
neurotic	impractical	submissive	compulsive

These are all ambiguous traits so it's hard to know how much of them you or anyone else really possesses. Note also that some of them are positive (the ones on the top row), while others are negative (the ones on the bottom row). When students at Cornell University answered the same questions relative to their Cornell peers, they said they expressed more of the positive traits than 64 percent of other Cornell students, but more of the negative traits than only 38 percent of Cornell students. This rosy view captures how we generally see ourselves—and perhaps it means that we'll pay close attention to the positive feedback and ignore the negative feedback we get on Instagram.

But as much as we value ourselves, we're also very sensitive to negative feedback. Psychologists call this the "bad is stronger than good" principle, and it's very consistent across different experiences. If you're like most people, your instinct is to scroll to the negative reviews on Amazon, TripAdvisor, and Yelp, because nothing cements an opinion like sharp criticism. You're also more likely to remember bad events from your past, and to ruminate over old arguments longer than you bask in recent praise. Even people who had happy childhoods, when asked to recall their lives as kids, are more likely to remember the few memories that were bad rather than the many that were good.

There are so many photos on Instagram that you might expect users to shrug off negative feedback. People should pay less attention to the "likes" under one Instagram photo than to the photos displayed at a solo art show or passed around to friends. In truth, though, the spotlight seems to find us even when we're in a crowd. In 2000, a group of psychologists asked college students to walk into a room filled with other students while wearing a T-shirt featuring a photo of Barry Manilow. (An unnecessary pre-test confirmed that college students prefer not to wear a Barry Manilow shirt in public.) After a few minutes, an experimenter escorted the unlucky subjects from the room, and asked them to guess how many of their fellow students noticed the Barry Manilow shirt. Of course they had been preoccupied by the shirt the entire time, so they guessed that half the students in the room would recall the shirt; in truth, only one in five remembered seeing Barry Manilow's likeness. A dud photo that attracts only three likes on Instagram is a bit like a Barry Manilow shirt. It's embarrassing to its owner, who assumes that other users are staring and laughing, when in fact they're far more concerned with their own photos, or at least with the endless line of photos that come before and after the "Manilow" shot.

The sting of negative feedback is so potent that many users take hundreds of shots before posting. Apps like Facetune allow tech novices to airbrush away their flaws for "perfect skin; a perfect smile," the ability to reshape their faces and bodies, remove blemishes, and recolor gray hair. Essena O'Neill, a young Australian model, had half a million followers when she decided to reveal the truth behind her glamorous Instagram posts. O'Neill changed her account name to *Social Media Is Not Real Life*, deleted thousands of old photos, and edited the captions under others. One photo featured O'Neill on the beach in a bikini:

NOT REAL LIFE—took over 100 in similar poses trying to make my stomach look good. Would have hardly eaten that day. Would have yelled at my sister to keep taking them until I was somewhat proud of this. Yep so totally #goals.

Under another shot of O'Neill in a formal dress by a lake:

NOT REAL LIFE—I didn't pay for the dress, took countless photos trying to look hot for Instagram, the formal made me feel incredibly alone.

And a third "candid" shot of O'Neill wearing a bikini:

Edit real caption: This is what I like to call a perfectly contrived candid shot. Nothing is candid about this. While yes
going for a morning jog and ocean swim before school was
fun, I felt the strong desire to pose with my thighs just apart
#thighgap boobs pushed up #vsdoublepaddingtop and face
away because obviously my body is my most likeable asset.
Like this photo for my efforts to convince you that I'm really
really hot #celebrityconstruct.

O'Neill attracted some backlash. Former friends accused her
of "100 percent self-promotion," and others called her new campaign "a hoax." But tens of thousands of others praised her publicly. "Read her captions—this girl is a boss," said one. "Aah, so
good, love what she's doing," said another. O'Neill was voicing
publicly what thousands of Instagram users felt across the globe:
that the pressure to present perfection with every shot is relentless
and, for many people, unbearable. In her last post, O'Neill wrote,
"I've spent the majority of my teenage life being addicted to social
media, social approval, social status, and my physical appearance.
Social media is contrived images and edited clips ranked against
each other. It's a system based on social approval, likes, validation,
in views, success in followers. It's perfectly orchestrated self-
absorbed judgment."

I n October 2000, Jim Young told his friend James Hong that
he'd met a girl at a party. According to Young, the girl was "a
perfect ten." Young and Hong had grown up together, attended
high school and then Stanford together, and now Young's com-

ment inspired them to design a website together. "This was on a Monday," remembers Hong. "It wasn't meant to be a serious project. We were just fooling around. Jim sent something to me on Friday or Saturday, I played with it over the weekend, and then we launched it the following Monday. So it was about a week from the idea to launching something."

The site was the online embodiment of Young and Hong's conversation. At 2 P.M. on the day of the launch, they asked forty-two of their friends to visit a webpage featuring Hong's headshot and a rating scale from 1 to 10. "Be nice," Hong told his friends, who were asked to decide whether Hong was "hot or not." The site was that simple: visitors rated one headshot after another, from 1 (not) to 10 (hot). After each rating, the screen refreshed to include the same headshot's average rating on the site. That way they learned instantly whether their internal beauty scale matched the scale used by other people. Forty thousand people visited the site the day after it launched. Eight days later, it was attracting two million hits a day—all without help from Facebook, YouTube, Twitter, and Instagram, which weren't due online for another several years. Visitors weren't just rating photos; they were also uploading their own, curious to know whether the online universe considered them hot—or not.

The site, which Hong and Young called Hot or Not, wasn't just viral; it was also addictive. And it wasn't just addictive to the usual crop of adolescent males. "I was looking at the site, and my dad walked into the room," Hong recalled. "You have to understand that at this point I was supposed to be getting a job, so I just told him, 'Oh, that's something Jim's working on.'" Hong's dad was curious, so Hong showed him how the site worked. After a quick demonstration, Hong's dad took the mouse and began rating. Hong remembers, "It was bizarre, because the first person I ever saw get addicted to rating people on whether they were hot or not was my dad. You have to understand, my dad is this sixty-year-old Asian engineering PhD guy who, as far as I was concerned, was asexual—except when he had me, my brother, and my sister." Hong's dad wasn't alone; millions of users were spending long stretches of time on the site, even willing to wait thirty seconds between pictures, which were painfully slow to download for the first few months.

Hong and Young created the site on a lark, but online advertisers began approaching them with serious offers. The friends stood to earn thousands of dollars a day, but for one catch: some of the photos were pornographic, and the advertisers were only willing to work with sites that promised to sanitize their content. Hong's parents had just retired, so he awkwardly asked them to trawl the site for porn. They had both developed low-level addictions to Hot or Not, so they happily obliged. If nothing else, their son had given them an excuse to spend more time on the site. At first they kept up with most of the site's new content. "Hey, it's going well! It's fun to look at people," Hong's dad reported gamely. When Hong's dad began sharing some of the banned photos with his son, Hong decided that he needed to find some new raters. He couldn't bear to imagine his parents looking at porn all day.

Hong and Young had no trouble enlisting some of their users as moderators. Like Hong's parents, they were glad to have a reason to spend hours a day browsing the site. In time, Hot or Not morphed into a dating website—a precursor to Tinder and other online dating platforms that prioritized looks over personality. Users paid just six dollars to join the site—a number that Hong and Young chose because it matched the price of two beers in a bar in the Midwest. At its peak, the site was generating $4 million in revenue each year, of which 93 percent was profit. The overhead for their lean, unintentionally addictive start-up was incredibly low. Rumor had it that Hong and Young's early success inspired Mark Zuckerberg to create FaceMash, the face-rating site that paved the way for Facebook. In 2008 the pair sold the site for $20 million to a Russian tycoon who specialized in online dating websites.

When they designed Hot or Not, James Hong and Jim Young were smart to include the same feature that made Instagram so successful: an engine for social feedback. After each rating, users discovered how closely their impressions matched the impressions of thousands of other users. Sometimes they matched and sometimes they didn't, and both outcomes satisfied basic human motives: the need for social confirmation when they matched, and the need for individuality when they didn't. (Of course it didn't hurt that users were rating facial attractiveness rather than, say, the attractiveness of different landscapes. With our inbuilt drive to scan the horizon for potential mates and competitors, we're naturally interested in physical attractiveness.)

Social confirmation, or seeing the world as others see it, is a marker that you belong to a group of like-minded people. In evolutionary terms, group members tended to survive while loners were picked off, one by one, so discovering that you're a lot like other people is deeply reassuring. When people are deprived of these bonds, they experience a form of pain so severe that it's sometimes called "the social death penalty." It's also very long-lasting—just remembering a time when someone excluded you is enough to rekindle the same agony, and people often list cases of social exclusion among their darkest memories. Discovering that you see a face the same way as other people see it is a route to belonging; it confirms that other people share your version of reality. Social confirmation is brief, and we need fresh doses all the time. It was this desire for repeated confirmation that nudged Hot or Not's users to rate "just one more photo" over and over again. One user who went by the handle Manitou2121 created a series of morphs averaging all the faces that received similar ratings. He shared the morphs with other users so they could see whether their views matched the views of the average Hot or Not user.

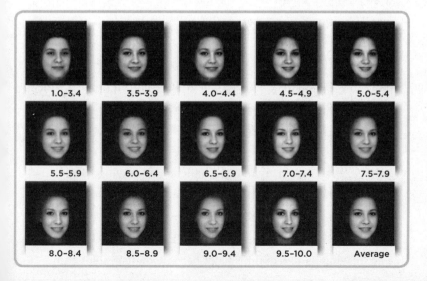

Occasional disagreement has its own benefits, though, because it serves to remind you that you're not like everyone else. Psychologists call this perfect balance the level of "optimal distinctiveness," and you tend to strike it when you agree with other people about most but not all things. Everyone strikes that balance differently but the beauty of Hot or Not was that it provided both forms of feedback. Hot or Not was the Instagram of photo-rating sites, but it could have just as easily gone the way of Hipstamatic had Hong and Young chosen to disable the feedback engine. Instead, it thrived as thousands of users were driven to discover whether their version of hot mirrored the version endorsed by everyone else.

I was just about to end my phone call with software engineer Ryan Petrie when he said, "It's interesting because I thought we were going to discuss my addiction to video games." Petrie grew up designing video games, so I'd called to ask him why some games are more addictive than others. I hadn't considered the possibility that he might be addicted to the very games he was designing. "I was very addicted for about eighteen months, when I was in college," Petrie told me. "I tried to spend all day, every day, online. I'd log in before class, between classes at the university library, and as soon as I got home after classes ended." On average, Petrie spent six to eight hours a day playing games, and "good" days were lost entirely to gaming. He flunked his classes, and spent a semester on academic probation. On the brink of expulsion, Petrie willed himself to spend more time in class and less time playing games, and his habit became manageable again.

Petrie is an old-school game designer. As a kid in the early 1980s, he watched his brother spend a summer programming a text-based *Wheel of Fortune* clone on an Apple IIe. To young Ryan this was magic. "My brother showed me a printout of the code, and I couldn't believe that this written incantation produced a video game. I asked him what each line did, over and over again, and soon I was making my own games." He began with a text-based Indiana Jones game that spanned three virtual rooms. He remembers it as "terrible," but soon he began to improve. EA Sports hired him after college, and more recently he's also spent time at Google and Microsoft.

"Have you heard of a MUD?" Petrie asked me. "A multiuser dungeon?" I hadn't, and based on its name I wasn't sure I wanted to know more. Petrie had been addicted to a MUD during college. MUDs are simple text-based role-playing games in which players type commands into the computer and watch as the screen refreshes with feedback and further instructions. Traditional MUDs feature scrolling text and no graphics, so they're capable of updating quickly even on very slow networks. They're completely free of the flashy sound and graphics that typify most of today's games, so all you're left with are the words on the screen and your imagination. Petrie's MUD of choice involved quests that he completed with other users from around the world. These users became his friends, and he felt guilty for abandoning them whenever he wasn't online. It was this social component of the game that kept Petrie hooked.

There's a certain purity to MUDs, because unlike modern games they don't rely on glitz and charm. Petrie was hooked entirely by the sense that he was playing alongside other people.

They may not have been in the room with him, but they all shared a common purpose. The MUD had a chat function, so players could commend each other on a job well done, or commiserate when they were defeated by powerful enemies. Petrie told me that MUDs still exist, but they've been swamped by big budget games—the showy Hollywood productions to his beloved indie masterpieces. "After all this time, that MUD is still the best game I've ever played. I always wanted to make one just like it, but after overcoming my addiction I questioned the morality of creating that sort of game."

Petrie's MUD was compelling, but it has nothing on today's most addictive games: massively multiplayer online games (or MMOs), like World of Warcraft or League of Legends. MUDs lived on the fringe, attracting a relatively small and sophisticated group of computer aficionados. In contrast, one hundred million people have opened WoW accounts. MMOs are more sophisticated than MUDs, but if you strip away their impressive graphics and sound effects, you're left with the same basic structure: a series of quests and remote interactions among gamers who become friends, relying on one another for support both within and beyond the game.

A couple of weeks after I spoke to Isaac Vaisberg, the former WoW addict I mentioned earlier, I visited reSTART's facilities in Washington State. Vaisberg obviously derived a lot of pleasure from his online friendships, so it wasn't clear to me why experts frowned on online interactions. Hilarie Cash, a clinical psychologist and cofounder of reSTART, explained that "there's nothing wrong with making friends online, as long as you also make friends in the real world. If we're good friends, and we're sitting together, that interaction, that energetic exchange releases a whole bouquet of neurochemicals that keeps us each regulated emotionally and physiologically. And it's our birthright as social animals to have lots of this sort of safe and caring interaction that keeps us regulated. We're not meant to be isolated islands." The addictive online friendships that attract young gamers are dangerous, not for what they provide, but for what they can't provide: a chance to learn what it means to sit, face-to-face, as you maintain a conversation with another person. The staccato taps of a keyboard—and even remote webcam interactions—obey a very different rhythm, and convey information along a much narrower bandwidth. "Even the smell of another person, the consistent eye contact that comes from being in the same room, is important," Cash said. She also reminded me that people who communicate by webcam never seem to look one another in the eyes, because the other person's eyes aren't perfectly aligned with the webcam that conveys your gaze. "It's a lot like feeding sugar to a hungry person," Cash told me. "It's pleasurable in the short-term, but eventually, they'll starve."

Cash invited me to participate in a group discussion session with the center's inpatients. As the session began, she repeated a mantra that I'd heard a couple of times already: "Remember: once your cucumber brain has become pickled, it can never go back to being a cucumber." The phrase was designed to discourage inpatients from doing what Vaisberg had done when he left the center: believing that they could play just one more game without their addictions returning. Cash was trying to explain that the inpatients' brains were forever pickled, in a sense, and that their addictions were always on the cusp of being rekindled. The mantra was a cute way of saying something very confronting: that it's impossible to ever completely escape the aftereffects of addiction. Cash also used the mantra to explain what happens when your brain is deprived of offline social interactions. As she told me, "If you only ever spend time online, a part of you withers away."

Cash suggested I speak to Andy Doan, a neuroscientist who had studied learning and memory at Johns Hopkins. She told me Doan was an expert on gaming addiction who could tell me more about the downsides of interacting with people online. I called Doan as soon as I returned to New York. He works as an eye surgeon now, but he has studied and written about addiction extensively. He told me that addictive games have three critical elements: "The first part is immersion—the sense that you're embedded in the game. The second is achievement—the sense that you're achieving something. And the third—and by far the most important—is the social element." Gaming addiction has risen dramatically, Doan said, because high-speed Internet connections have made it easier to communicate with other players in real time. Gone are the days of clunky networks and Ryan Petrie's beloved, but peripheral, MUDs, which addicted a much smaller set of people. Now Isaac Vaisberg and tens of millions of other gamers can build simulated friendships that almost look and feel like the real thing.

Doan explained why a brain raised on online friendships can never fully adjust to interactions in the real world. In the 1950s and 1970s, in a famous series of experiments, vision researchers Colin Blakemore and Grahame Cooper showed that what a young kitten sees shapes how his brain works for the rest of his life. In one experiment, they confined the kittens to a very dark room until they were five months old. Once a day, they removed half the kittens from the room and placed them in a cylinder covered with horizontal black and white stripes. They removed the other half and placed them in a similar cylinder, this one covered with vertical black and white stripes. So, half the kittens saw only vertical lines, and half saw only horizontal lines. They explained that, for each kitten, "There were no corners to its environment, and the upper and lower limits to its world were a long way away. It could not even see its own body, for it wore a wide black collar that restricted its visual field." They added, providing little comfort to anyone even remotely concerned with animal welfare, that "The kittens did not seem upset by the monotony of their surroundings and they sat for long periods inspecting the walls of the tube."

When Blakemore and Cooper allowed the kittens to roam a normal room, they were very confused. All of them, regardless of whether they'd been exposed to horizontal or vertical lines, struggled to judge how far away they were from physical objects. They bumped into table legs, failed to jump back when the experimenter acted like he was about to tap their faces, and couldn't follow moving objects unless they made a noise. (If you've seen how avidly cats follow laser pointers, you know how strange it is when a cat ignores a rolling ball.) When Blakemore and Cooper examined the kittens' brains for activity, they found that the kittens reared in vertical environments showed no activity at all in response to horizontal lines, while those reared in horizontal environments did not respond to vertical lines. Their brains were

effectively blind to whatever they hadn't been exposed to naturally during the first few months of their lives. This, Andy Doan
told me, was irreversible. The visual cortex inside these poor kittens' heads had been pickled forever, and even exposing them to
normal environments for the rest of their lives did nothing to
reverse many of the effects of their stunted early lives.

Doan drew an analogy to Hilarie Cash's reSTART inpatients. The technical term for what Blakemore and Cooper in-
duced in their kittens is visual amblyopia (Greek for "blunt
vision"). Doan told me that children reared on the Internet suffer
a kind of emotional amblyopia. Children develop different mental skills at different ages, during so-called critical periods. They
pick up new languages with ease until ages four or five, after
which they only pick up new languages with considerable effort.
A similar idea holds for developing social skills—and for learning how to navigate the complex world of teenage sexuality. If
kids miss out on the chance to interact face-to-face, there's a fair
chance they'll never acquire those skills.

Cash has seen dozens of adolescents, mainly boys but also
girls, who have no problem interacting with peers online, but
can't carry a conversation with someone sitting across from them.
The problem worsens when you encourage adolescent males and
females to interact. "How do you learn to talk and flirt and date
and end up in bed if you've only mixed with other people online?" Cash asked. "Our guys get sidetracked, and they develop
intimacy disorders. They don't have the skills to bring sexuality
and intimacy together. Many of them turn to pornography instead of forming real relationships, and they never seem to understand true intimacy." Cash referred to "our guys" because the

center no longer admits women. "For four years we admitted women, but we had to revise our policy after a number of patients ignored the 'no physical intimacy' rule. We had many more male applicants in those days, so we decided to stop taking women. Now, with the rise of non-violent casual and social gaming, there are almost as many female applicants. We may have to reconsider our policy."

Even addicts who, like Isaac Vaisberg, somehow win the charisma lottery are susceptible to a range of psychological and social disorders. One study found that gamers aged between ten and fifteen years who played more than three hours per day were less satisfied with their lives, less likely to feel empathy toward other people, and less likely to know how to deal with their emotions appropriately. Three hours may sound like a lot, but recent surveys have shown that kids spend an average of five to seven hours in front of screens each day. When today's Millennials become adults, there's a fair chance their social cucumber brains will be pickled.

PART 3

The Future of
Behavioral Addiction
(and Some Solutions)

10.

Nipping
Addictions at Birth

Today, the average schoolchild aged between eight and eighteen years spends a third of her life sleeping, a third at school, and a third engrossed in new media, from smartphones and tablets to TVs and laptops. She spends more time communicating through screens than she does with other people directly, face-to-face. Since the turn of the new millennium, the rate of non-screen playtime fell 20 percent, while the rate of screen playtime increased by a similar amount. These stats aren't inherently bad—the world changes constantly—but in 2012 six researchers showed that they happen to be taking a human toll.

In the summer of 2012, fifty-one children visited a summer camp just outside Los Angeles. The children were typical Southern Californian public school kids: an equal mix of boys and girls

aged eleven or twelve years from a variety of ethnic and socioeco-
nomic backgrounds. All of them had access to a computer at
home, and roughly half owned a phone. They spent an hour texting friends each day, about two and a half hours watching TV,
and just over an hour playing computer games.

For this one week, the children would leave their phones and
TVs and gaming consoles at home. Instead, they hiked and
learned to use compasses and to shoot bows and arrows. They
learned how to cook over a campfire and how to tell an edible
plant from a poisonous plant. They weren't explicitly taught to
look each other in the eyes, face-to-face, but in the absence of new
media, that's exactly what happened. Instead of reading "LOL"
and staring at smiley-face emojis, they actually laughed and
smiled. Or didn't laugh and smile if they were sad or angry.

On Monday morning, when the kids arrived at the camp,
they took a short test called the DANVA2, which stands for the
Diagnostic Analysis of Nonverbal Behavior. It's a fun test—one
of those tests that goes viral on Facebook—because all you have
to do is interpret the emotional states of a bunch of strangers.
For half the test you look at their faces in photos, and for the
other half you listen to them read a sentence aloud. Then you
decide whether they're happy or sad or angry or fearful. That
may sound trivial, but it isn't. Some of the faces and voices are
easy to read—these are labeled "high-intensity"—but many of
them are subtle. Like deciding whether the Mona Lisa is smiling inside, or whether she's just bored or unhappy. I tried the test
and got some of the answers wrong. One guy sounded mildly
depressed, but the test told me he was actually mildly afraid. The

summer camp kids had the same experience. They made an average of fourteen errors across the forty-eight-item test.

Four days of camping and hiking later, the kids were ready to file onto buses to return home. Before they did, the researchers administered the DANVA2 again. They reasoned that a week of face-to-face interaction without distraction from gadgets might make the kids more sensitive to emotional cues. There's good reason to believe practice makes perfect when it comes to reading emotional cues. Children who are raised in isolation—like the famed Wild Boy of Aveyron who was raised by wolves in a forest in France till he was nine years old—never learn to read emotional cues. And people who are forced into isolation struggle to interact with others when they emerge, sometimes for the rest of their lives. Children who spend time together also learn to read emotional cues through repeated feedback: you may think your playmate is holding out a toy because he wants to share it with you, but if you look at his face you'll see he's about to use the toy as a weapon.

Reading emotions is a finely tuned skill that atrophies with disuse and improves with practice, and that's what the researchers found at the summer camp. The kids did much better the second time they took the DANVA2. They were never told the answers to the test after taking it the first time, but their error rate dropped by 33 percent. The researchers also asked a control group of kids from the same school to take the test twice. These kids didn't attend the camp, so they took the test on a Monday morning and a Friday afternoon just as the camp kids did. Their error rate dropped a bit, too—by 20 percent—presumably because there's some benefit to taking the same test twice, but this rate of improvement was much less impressive than the rate shown by the wilderness campers.

Now, there's a lot that separates a week in the city from a week at camp. Apart from access to gadgets and time spent face-to-face with friends, there are plenty of other differences that may have explained the kids' different rates of improvement on the DANVA2. Is it that spending time in nature improves mental functioning? Or that spending time with your peers makes you smarter? Or that staying away from gadgets makes all the difference? It's impossible to be sure, but that doesn't change the prescription: kids do better at a task that drives the quality of their social interactions when they spend more time with other kids in a natural environment than they do when spending a third of their lives glued to glowing screens.

Children are especially vulnerable to addiction, because they lack the self-control that prevents many adults from developing addictive habits. Regulated societies respond by refusing to sell alcohol and cigarettes to children—but very few societies regulate behavioral addictions. Kids can still play with interactive tech for hours at a time, and they can still play video games as long as their parents will allow. (Korea and China have flirted with so-called Cinderella laws, which prohibit children from playing games between midnight and six in the morning.)

Why shouldn't kids be allowed to play with interactive tech for hours at a time? And why, as I mentioned in the book's prologue, do so many tech experts prohibit their children from using the very devices they design and promote in public? The truth is that we won't know how children will respond to tech overuse for some years still. The first generation of native iPhone users is only eight or nine years old, and the first generation of native iPad users is six or seven. They haven't reached their teens, so there's no way to know just how different they are from their peers who are just a couple of years older. But we do know what to look for. Tech subsumes some very basic mental activities that were once universal. Kids of the 1990s and earlier stored dozens of phone numbers in their heads; they interacted with each other rather than with devices; and they made their own fun instead of extracting manufactured fun from ninety-nine cent apps.

A couple of years ago, I became interested in what we call *hardship inoculation*. This is the idea that struggling with a mental puzzle—trying to remember a phone number or deciding what to do on a long Sunday afternoon—inoculates you against future mental hardships just as vaccinations inoculate you against illness. Reading a book, for example, is harder than watching the TV. (David Denby, a film critic at the *New Yorker*, recently wrote that kids are abandoning books as they age. "Books smell like old people," he overheard one teenager say.) There is good early evidence to support the idea that small doses of mental hardship are good for us. Young adults do much better on tricky mental puzzles when they've solved difficult (rather than easy) ones earlier. Adolescent athletes also thrive on challenges: we've found, for example, that college basketball teams do better when their pre-season schedules are more demanding. These mild initial struggles are critical. Depriving our kids of them by handing them a device that makes everything easier is dangerous—we just don't know how dangerous.

Relying too heavily on tech also leads to a phenomenon known as *digital amnesia*. In two surveys, thousands of U.S. and European adults struggled to remember a raft of important phone numbers. They struggled to recall their kids' cell phone numbers and the central phone number at their workplace. In other questions, 91 percent of respondents described their phones as "extensions of their brains." The majority said they would search online for answers to questions before trying to generate the answer from memory, and 70 percent said they would feel sadness or panic if they lost their smartphone devices even briefly. Most said there was information on their smartphones that wasn't stored in their heads or anywhere else.

The M.I.T. psychologist Sherry Turkle has also argued that technology turns children into poor communicators. Take the case of texting, which many children (and adults) prefer to phone calls. Texting allows you to modulate your message more precisely than does speech. If you usually reply "haha" to a joke, you can write "hahaha" to signal that this one is particularly funny—or "HAHAHAHA" if the joke is uproarious. If you're angry, you can reply with the dismissive "k," and if you're furious you can choose not to respond at all. To shout, use a single "!," and to exclaim loudly, use "!!" or even "!!!!" There's a mathematical precision to these signals—you can count the number of "ha"s or "!"s—so texting is ideal for the risk-averse communicator who worries about miscommunicating. The significant downside is that nothing is spontaneous and very little is ambiguous when you follow the rules of text-speak. There are no non-verbal cues; no pauses and lilts and unplanned giggles or scoffs to punctuate your partner's message. Without these cues, children can't learn to communicate face-to-face.

Turkle illustrates the limitations of cell phone communication by recounting an observation that comedian Louis C.K. shared with Conan O'Brien in 2013. He explained that he was not raising children; he was raising the grown-ups they're going to be. Phones, he said, are "toxic, especially for kids."

> They don't look at people when they talk to them. And they don't build the empathy. You know, kids are mean. And it's because they're trying it out. They look at a kid and they go, "You're fat." And they see the kid's face scrunch up and they go, "Ooh, that doesn't feel good to make a person do that" . . . but when they write, "You're fat," then they just go, "Mmm, that was fun. I like that."

For Louis C.K., face-to-face communication is essential, because it's the only way for kids to appreciate how their words affect other people.

As I write this, two weeks ago my wife gave birth to our first child. Sam Alter was born into a world of screens. The screen on his baby monitor carries our voices and faces to his room. The screen on my iPad introduces him to his grandparents and uncle and cousins around the world. The TV in our living room delivers moving pictures and sounds as we soothe him to sleep. In time he'll learn to use the iPad and the TV himself. Then he'll learn to use computers and smartphones, and whichever soon-to-be-invented devices come to define his generation as computers and smartphones define ours. In many respects these screens will enrich his childhood: he'll watch videos, play games, and interact with people in ways his ancestors considered science fiction. But there's a good chance they'll also detract from his childhood. Two-dimensional screen worlds are poorer versions of the real thing. Social interactions are watered down, and there's more room for spoon-feeding and less room for imagination and exploration. As Andy Doan told me, the time we spend with screens as children goes on to influence how we interact with the world for the rest of our lives. It's easier to strike the right balance earlier than to correct unhealthy patterns later.

One subgenre of YouTube videos shows how infants respond to screen time: they don't know how to use magazines. One of these videos has over five million views. It shows a one-year-old girl who swipes an iPad screen like a pro. She navigates from one screen to the next and squeals happily as the device responds to her will. The swiping gesture that Apple introduced with its first iPhone in 2007 is as natural to her as breathing or eating. But when she sits in front of a magazine, she continues to swipe, becoming frustrated when the inert photos before her refuse to resolve into new ones. She is among the first humans to understand the world this way—to believe that she has limitless command over the visual environment, and the ability to overcome the staleness of any experience by welcoming its replacement with a dismissive swipe. The video is aptly titled, "A magazine is an iPad that does not work," and the comments below the video ask questions like, "Can you explain why you'd give a one-year-old an iPad?"

iPads make the job of parenting much easier. They provide renewable entertainment to kids who like watching videos or playing games, so they're a miracle for overworked and under-rested parents. But they also set dangerous precedents that are difficult for kids to shake as they mature. reSTART's Hilarie Cash has firm views on the subject. She isn't puritanical, but she's seen the effects of overexposure firsthand. "Kids shouldn't be exposed to screens before the age of two," she says. Their inter-actions, Cash argues, should be direct, social, firsthand, and concrete. Those first two years set the standard for how those kids will interact with the world when they're three and four and seven and twelve years old, and so on. "They should be al-lowed to watch passive TV till they reach elementary school—around age seven—when they should be introduced to interactive media, like iPads and smartphones," Cash says. She also suggests limiting screen time to two hours per day, even for teenagers. "It's not easy," she admits. "But it's critical. Kids need sleep and physi-cal activity, and family time, and time to use their imaginations." Those things can't happen when they're lost in screenworlds.

The American Academy of Pediatrics (AAP) agrees with Cash. "Television and other entertainment media should be avoided for infants and children under age 2," the AAP advised in an online report. "A child's brain develops rapidly during these first years, and young children learn best by interacting with peo-ple, not screens." That may be true, but abstinence is a lot to de-mand when screens are everywhere. Even in 2006—four years before Apple introduced the first iPad—the Kaiser Foundation found that 43 percent of children under two watched TV daily, and 85 percent watched at least once a week. Sixty-one percent of children under two spent at least some time each day in front of a screen. In 2014, an organization called Zero to Three reported that 38 percent of children under two years of age had used a mobile device (compared to 10 percent in 2012). By age four, 80 percent of children had used a mobile device.

Zero to Three takes a softer approach than the AAP, acknowledging that some screen time is all but inevitable. Rather than banning screens outright, Zero to Three recommends specific *kinds* of screen time. Its report begins:

> A robust body of research shows that the most important factor in a child's healthy development is a positive parent-child relationship, characterized by warm, loving interactions in which parents and other caregivers sensitively respond to their child's cues and provide age-appropriate activities that nurture curiosity and learning.

The AAP obviously agrees: its statement on infant media consumption ends with the words "young children learn best by interacting with people, not screens." The difference between them is that Zero to Three acknowledges that children can develop healthy interactions with screens, as long as parents are involved as well. Instead of banning screens, it lists three major qualities of healthy screen time.

First, parents should encourage their children to connect what they see in the screen world to their experience of the real world. If an app asks children to sort wooden blocks by color, parents might ask those children to label the color of clothes as they sort the laundry together. If an app presents wooden blocks and balls, children should play with real wooden blocks and balls afterward. No experience should be confined to a virtual world that was designed to mimic reality. This screen-to-reality bridging is known as *transfer of learning*, and it improves learning for two reasons: it requires children to repeat what they've learned, and it encourages them to generalize what they've learned beyond a single situation. If a dog on the screen is the same as a dog on the street, the child learns that dogs can exist in many contexts.

Second, active engagement is better than passive viewing. An app that requires children to act, remember, decide, and communicate with their parents is better than a TV show that allows them to absorb content passively. Slower-paced shows, like *Sesame Street*, encourage participation and engagement, so they're superior to faster-paced shows like *SpongeBob SquarePants* (which isn't designed for children under five). In one study, four-year-olds who watched *SpongeBob* (rather than a slower-paced educational cartoon) for nine minutes struggled to remember new information and resist temptation afterward. Consequently, the TV should rarely be on in the background, and TV time should be separated from the rest of the day.

Third, screen time should always focus on the content of the app rather than the technology itself. Children who are watching a story unfold should explain what they think might happen next; to point to and identify the characters on the screen; and to move slowly enough through the process that they aren't overwhelmed by navigating the technology. To the extent possible, a screen-based story should mimic the experience of a book.

Like younger children, adolescents tend to be vulnerable to addiction. reSTART uses the metaphors of dieting and environmental sustainability to describe when and how often older children should interact with screens. Cash told me that she preferred not to use the term "addiction," which implies all the trappings of disease. Instead, the center adopts the language of the environmental movement. Its homepage declares that reSTART is a "Center for Digital Technology Sustainability," and that it teaches people to live sustainable lifestyles. The center is a "retreat" rather than a treatment facility. "It's impossible to avoid technology altogether, so our aim isn't to teach our clients to go cold turkey," Cash told me. "We teach people how to problem solve, which isn't a traditional form of therapy." Cash explained that problem solving was critical, because the treatment plan lasted only forty-five days. After that, the boys were on their own.

reSTART's treatment plan has three phases. During the first phase, the patients aren't allowed to use technology at all. They go through detox, which typically lasts around three weeks. "Some of the guys are very resistant, but others embrace the process," Cash said. "We can usually tell by the end of the first phase who will benefit from the treatment, and most of them do." For the remainder of the first phase—another three or four weeks—the boys continue to live at the center. They learn the basic life skills that many of them lack, like cooking an egg, cleaning a toilet, making their beds, and, most important, managing their emotions. (One of the boys told me that he had played several games of chess since joining reSTART, and they usually ended when he threw the loaded chess board in anger.) They also learn to exercise and to embrace nature, which is a big part of reSTART's philosophy: if you're going to strip away a major component of their lives, you have to replace it with something that engages them and allows them to escape technology. Cash's cofounder, Cosette Rae, told me that her husband leads nature walks. reSTART is nestled in a large forest, but beyond that the boys also hike nearby Mount Rainier. They train at the center's gym every day, and many of them become quite fit. Cash cited an independent study, which found that 78–85 percent of the boys improve over this initial phase.

During the second phase, the recovering patients move into halfway houses that are similar to the houses managed by Alcoholics Anonymous. There they learn to apply the skills they learned at reSTART. They apply for jobs or volunteer positions, or they take college classes. The houses run according to stringent rules, and the patients receive support from reSTART in exchange for showing up at the center for regular outpatient check-in appointments. I asked Cash whether the program had been successful, and she told me that it had, but that she couldn't provide firm numbers. reSTART is small, and the nature of each boy's problem is slightly different, so it's difficult to measure relapse rates. A graduate student was in the process of working with Cash and Rae to implement a more rigorous measurement plan.

The third and final phase begins when former patients are ready to return to life without supervision. Many of them stay in Washington, near the center, which allows them to check in with the center every few weeks or months. Since they come from all over the country, and occasionally from outside the United States, they're also less likely to be tempted by old habits if they avoid the people and places that characterized their former addictions. (Remember the Vietnam vets who escaped their heroin addictions when they returned home from Vietnam.) Isaac Vaisberg discovered this the hard way when he returned home after his first visit to reSTART, and couldn't resist playing World of Warcraft. He decided to stay nearby after leaving the center for the second time, and he still lives a short drive from reSTART.

Most teens don't need to spend time at a facility like reSTART, but their parents still worry about how they interact with games and social media. Catherine Steiner-Adair, the psychologist I mentioned in chapter 1, interviewed thousands of teens and their parents to formulate a set of basic parenting principles. The teens, she explained, react badly to parents who are "Scary, Crazy, and Clueless."

Scariness comes in the form of rigid, judgmental intensity. As parents become more worried, their claims naturally escalate. Statements like, "You'll ruin your chances for college!" or "You will never bring that friend into this house again!" are guaranteed to alienate kids. Crazy parents overreact when their children come to them with problems. Steiner-Adair describes the case of a twelve-year-old girl who received a hurtful email from her friend. "She couldn't talk about things like this because mom always had a way of ramping up the drama on everything. 'She'll say "that's horrible!" and then get started, and then I not only have my friend to deal with but my crazy mother, too.'" It's clear that the girl's mother cares—she wants her daughter to feel better—but her instinctive, escalating response makes the problem worse. Clueless parents, on the other hand, are objects of pity. They either don't understand the lives their children lead, or they find it overwhelming. "A clueless parent tries too hard" to be-friend his kid, says Steiner-Adair. "He misses cues, often engaging over superficial things while failing to have meaningful conversations with his child about life values and about expectations and consequences."

In contrast to the scary, crazy, and clueless parents are those who are "Approachable, Calm, Informed, and Realistic." They understand that social media is a part of the real world. Sometimes their children will be upset, but overreaction makes the problem worse. These parents take the time to understand how their kids interact with social networking platforms. They ask non-judgmental questions of their kids and do their own research. They also impose boundaries, creating the sort of sustainable relationship with tech promoted at reSTART. The family engages in meaningful offline conversations, and at certain points in the day, everyone is offline together. Some of these ideals might seem obvious in the abstract, but they're not always easy to achieve in the heat of the moment. Steiner-Adair's mantra—Approachable, Calm, Informed, Realistic—is a useful rule of thumb when tensions rise.

So far, the U.S. government has chosen not to intervene on the relationship between children and behavioral addiction. There are no state-sponsored treatment clinics, perhaps because a relatively small percentage of addicted children need psychiatric help. The East Asian response to behavioral addiction, particularly in China and South Korea, has been far more adventurous than the U.S. response. In 2013, two Israeli filmmakers released a documentary called *Web Junkie*. For four months, Hilla Medalia and Shosh Shlam interviewed doctors, patients, and parents at an Internet addiction treatment facility in Beijing, China. Several years earlier, China had become the first country to declare Internet addiction a clinical disorder, labeling it "the number one public health threat" to its teenage population.

There are more than four hundred treatment centers in China, and, according to the country's definition of Internet addiction, more than twenty-four million teen Internet addicts. Medalia and Shlam visit one of these centers, the Daxing Camp at Beijing Military Hospital, where they speak to the doyen of Internet addiction treatment in China, Professor Tao Ran. Ran is a soft-spoken psychiatrist who nonetheless inspires hatred in the center's patients. Most of them have been tricked into visiting the center, where they spend three or four months as involuntary inpatients. They are forced to take a regimen of pills and to march in military formation, even as temperatures plunge in the midst of the harsh Beijing winter. Their parents, many of whom weep openly on camera, commit their sons (and occasionally daughters) because they feel they have no other option. Early in the documentary, Professor Ran explains the problem and his role as the center's director:

> Internet addiction is a cultural problem among Chinese teenagers. It has surpassed any other problem. As a psychiatrist, my job is to determine if this is a disease. We notice that these children have a bias towards virtual reality. They think the real world isn't as good as the virtual world. Our research shows that addicts spend more than six hours a day online,

not for the purpose of work or studying . . . some kids are so hooked on these games that they think going to the bathroom will affect their performance. So they wear a diaper. They are the same as heroin addicts—they crave and look forward to playing every day. That's why they call it "electronic heroin."

Later, Professor Ran implies that the problem is structural—that it isn't a disease, and that it lies with society. He meets with a group of parents in a small, depressing room at the center. "One of the biggest problems among these kids is loneliness. Loneliness. Did you know they feel lonely?" he asks, speaking through an oddly echoing microphone that seems better suited to an arena. One parent replies, "I think it's because they're the only child in the family. And as parents we fail to make friends with our children. We only ask them to study hard. Their stress, their worries, their pain—we can't see any of it. We care only about their studies." Ran agrees. "So where do they look for their friends? The Internet. The virtual world has all sorts of grand audio and visual extravaganzas. Simulations that you won't be able to find anywhere else. It becomes their best friend." It's clear that Professor Ran is ambivalent about the nature of Internet addiction. On the one hand he forces his patients to take psychotropic drugs, while on the other he implies that it isn't a disease at all. When a society churns out millions of lonely, overworked children, why wouldn't they turn to a boundless source of companionship and escape? That seems like a rational response to their disaffection. What brings about their undoing isn't that they're suffering from a disease, but rather that this digital world is so clearly superior to the real world they're supposed to be inhabiting instead.

The teens themselves recognize this. They're sophisticated in ways that escape the adults who grew up in a relatively primitive world. A group of teenage boys at the center discusses their addiction in a show of masculine one-upmanship. One says he played a video game for two months straight without quitting—the full extent of his summer vacation. Another chimes in and says he played for three hundred days, stopping briefly to eat, sleep, and use the bathroom. A third calls Professor Ran's definition of addiction "bullshit." Six hours a day seems normal to him. "If you check their definition of Internet addiction, 80 percent of Chinese must have it." A fourth says, "Most of us don't think we have Internet addictions. It's not a real disease. It's a social phenomenon." The boys try to minimize the issue, but it's clear that Internet addiction is a massive and growing problem in China.

The Western approach to behavioral addiction is just as scattered as Professor Ran's approach. The *Diagnostic and Statistical Manual* now recognizes that gambling is a genuine behavioral addiction, and excessive Internet use was almost included in the DSM's fifth edition, published in 2013. There are now more than two hundred academic papers on the topic of "Internet addiction," so the American Psychiatric Association chose to mention it briefly in the manual's sppendix. Meanwhile, the DSM omitted other behavioral addictions, like exercise, smartphone, and work addiction, because they hadn't yet attracted enough academic interest. That doesn't make the experience of these addictions any less real, though, as I discovered when I spoke to behavioral addiction treatment experts. Even if the APA doesn't consider them diseases or disorders, they still affect many thousands of lives. And perhaps they shouldn't be considered clinical disorders at all—perhaps, like the millions of Chinese teens who treat loneliness by turning to the Internet, behavioral addicts are just responding to the constraints of the world they happen to inhabit.

In contrast to Professor Ran's medical model, with its pills and psychiatric treatment sessions, reSTART primarily treats behavioral addiction as a structural issue: fix the structure of the affected person's life and you'll fix the problem. Therapy sessions form a small part of reSTART's treatment plan—far smaller than, say, life training and coping skills. But that isn't true of every U.S. facility. There is one hospital that treats behavioral addiction much as Western medicine treats substance addiction. The Bradford Regional Medical Center in Pennsylvania launched a ten-day inpatient treatment program for Internet addicts in 2013. Kimberly Young, the psychologist who founded the program, became interested in Internet addiction in the mid-1990s. "In 1994 or 1995, a friend of mine told me that her husband was spending between forty and sixty hours a week in AOL chat rooms," Young said. "Internet access was expensive then, at $2.95 per hour, so his habit became a financial burden. I wondered whether people could develop an addiction to the Internet." Young created the Internet Addiction Diagnostic Questionnaire, or IADQ, which she posted online. Like gambling and alcohol addiction questionnaires, the IADQ asked respondents to indicate whether eight statements applied to them. "Anyone who said

at least five of the statements applied to them was 'addicted,'"
Young told me.

The next day dozens of people emailed her to say they were
concerned. Many of them were scoring above five on the scale.
Over the next four years, Young refined and validated the questionnaire, added twelve new items, and renamed it the Internet
Addiction Test. (I included a sample of questions from the test in
the first chapter of this book.)

Young began to treat a growing list of Internet addicts, fueled
by two specific events, first in 2007 and then in 2010: the introduction of Apple's iPhone and then its iPad. "My focus on Internet addiction exploded when the Internet went mobile," Young
told me. The addiction context was no longer limited to the
home—now it was everywhere. By 2010, Young recognized the
need for a dedicated treatment center. A long-out-of-date study
in 2006 suggested that one in eight Americans was addicted to
the Internet, but Young was convinced the number was much
higher—and rising. She managed to secure sixteen beds at Bradford, which were set aside for an acute Internet addiction treatment center. She had spoken to Cash at reSTART, but Young
preferred a different, more intensive approach. Instead of forty-
five days, patients would stay at her facility for just ten days.
"Most people don't have time to spend longer than ten days with
us," she said. Many of them had seen other doctors who couldn't
help, so by the time they arrived at the hospital, they were desperate. They would undergo a rapid three-day detox, and then seven
days of targeted cognitive-behavioral therapy. Young's approach,
known as Cognitive Behavioral Therapy for Internet Addiction,

or CBT-IA, borrowed techniques that had been successful in treating other impulse disorders. Many of her patients don't believe they have a problem, so she has to teach them to recognize that they are, in fact, addicts. Then she teaches them to reframe some of the harmful ideas that lead them to overuse the Internet— for example, the notion that they're incapable of forming friendships offline. CBT-IA also encourages patients to re-engage with the offline world, which many of them have abandoned in favor of an online world that seems more forgiving.

In 2013, Young published a paper that described the effects of CBT-IA on 128 Internet addicts. She measured their progress immediately after twelve treatment sessions, and again one month, three months, and six months after treatment ended. The results were encouraging: immediately after treatment, Young's patients were less preoccupied with the Internet, more capable of managing their time, and less likely to be experiencing harmful consequences from overuse. Six months later some of the treatment benefits had weakened, but the patterns were similar: CBT-IA seemed to be working, at least on this limited sample.

Programs like reSTART, Kimberly Young's CBT-IA, and Professor Ran's military academy are desperate attempts to deal with the most severe cases of behavioral addiction—and they're restricted to Internet and gaming addiction. They aren't perfect, but early evidence suggests that they yield small to moderate benefits. But what are we supposed to do with the remaining millions who aren't ready or able to be inpatients—the millions who exercise too often, work longer hours than they should, and can't help spending too much money online?

The answer is not to medicalize these moderate forms of addiction, but to alter the structure of how we live, both at a societal level and more narrowly, as we construct our day-to-day lives. It's far easier to prevent people from developing addictions in the first place than it is to correct existing bad habits, so these changes should begin not with adults, but with young kids. Parents have always taught their children how to eat, when to sleep, and how to interact with other people, but parenting today is incomplete without lessons on how to interact with technology, and for how long each day.

Like Alcoholics Anonymous, many clinical programs promote abstinence: either you abstain from the addictive behavior, or you'll never shake the addiction. Since abstinence isn't a practical option for many modern behaviors, one alternative intervention takes a different approach. Where Alcoholics Anonymous suggests that addicts are helpless to overcome their addictions, *motivational interviewing* rests on the idea that people are more likely to stick to their goals if they're both intrinsically motivated and feel empowered to succeed. Counselors begin by asking open-ended questions that encourage their clients to consider whether they want to change their addictive behaviors. What makes the approach radical is that clients are allowed to decide they don't want to change their behavior at all.

Carrie Wilkens, cofounder and clinical director of the Center for Motivation and Change in New York City, explained the process. "The key to motivational interviewing is getting the costs and also the benefits of the addictive behavior on the table. We all know how terrible addiction is, but it also has benefits, and this tends to be the most meaningful part of the puzzle. Unpacking the behavior's benefits is great because then you can understand the underlying needs that the behavior addresses."

If, for example, a sixteen-year-old girl checks her Instagram account dozens of times a day, she might say that the benefit is that she feels connected to her friends. She posts pictures three or four times a day, and feels compelled to check whether her posts are attracting likes. The key to treating her addiction, then, is to make sure she feels connected through other means, and that she feels validated in the absence of those likes. A typical session with the girl might begin with something called a *readiness ruler*:

On a 0–10 scale, if 0 is not in the least bit ready to change your behavior, and 10 is as fired up as you can be, where are you?

The first question in the primary intervention probes the girl's response to this question. Why is the number so high or low? This gives her a chance to express her willingness to change. If she gives a low response, she might say she doesn't see any need to change her behavior; with a high response she might admit that her Instagram use is making her unhappy. From there, the clinician asks a series of open-ended questions:

What are the benefits of your Instagram use?
How would you like things to be different?
How does your Instagram use affect your well-being?
In what ways do you feel you could be doing better?

Counselors who practice motivational interviewing complete rigorous training seminars, but the general approach has plenty of benefits for parents and even adults who are trying to change their own behavior. It's non-judgmental by nature, so addicts are less likely to be defensive. One script, for example, suggests the following opening:

> I'm not here to preach to you or tell you what you "should" do; how would I know, it's your life and not mine! I believe people know what's best for them.
>
> I don't have an agenda, just a goal: to see if there is anything about the way you take care of your health that you would like to change, and if so, to see if I can help you get there.
>
> How does that sound to you?

Counselors traditionally used the approach to treat substance abuse, but Wilkens says it works just as well for behaviors. At least one study confirmed her belief. It works because it motivates people to change, and gives them a sense of ownership over the process. They aren't being cajoled or pressured to change by someone else; they're choosing to change voluntarily. The approach also recognizes that different people are driven to overcome their addictions by different motives. For some people, addictions are a barrier to productivity; for others, a barrier to health; and for many, a barrier to fulfilling social relationships. Motivational interviewing uncovers that motive, and prompts the addicted person to change.

The technique's effectiveness is explained by one of the dominant theories in motivation research: Self-Determination The-ory (SDT). SDT explains that people are naturally proactive, especially when a behavior activates one of three central human needs: the need to feel in command of one's life (autonomy); the need to form solid social bonds with family and friends (relat-edness); and the need to feel effective when dealing with the external environment—learning new skills and overcoming chal-lenges (competence). Though addictive behaviors are designed to soothe psychological discomfort, they also tend to frustrate one or more of these needs. A motivational interview makes that frus-tration clear: if you're asked how your Instagram use affects your well-being, you're going to see that it's compromising your pro-ductivity, your relationships, or both. Far from rendering a per-son powerless in the face of her addiction, she's left to feel both motivated and capable of changing for the better.

SDT emerged during the extravagant mid-1980s. Wall Street excess had reached a peak, and businesses believed that workers responded best to bigger paychecks and lavish perks. SDT sug-gested that these forms of compensation—known as extrinsic rewards—would fail to sustain motivation in the long run. What workers needed were intrinsic rewards: a job that made them feel effective and competent at a company they respected. Sometimes, extrinsic rewards were actually counterproductive, because they robbed workers of genuine intrinsic motivation. In one experi-ment, students enjoyed completing a series of puzzles—until re-searchers started paying them. As soon as they were paid, the students decided the puzzles weren't much fun after all. When given the chance to continue working on the puzzles, they pre-ferred other activities instead. SDT shows how important it is to design the right sort of environment, regardless of whether you want to promote or discourage a behavior. The key is to understand how different features of the environment—financial incentives and physical barriers, for example—shape motivation. A well-designed environment encourages good habits and healthy behavior; the wrong environment brings excess and—at the extremes—behavioral addiction.

11.

Habits and Architecture

In the United States, politics and religion go hand-in-hand. Conservative states tend to be religious, and liberal states tend to be secular. That first category includes Mississippi, Alabama, Louisiana, South Carolina, and Arkansas. All five are Southern states that fall within the Bible Belt—the epicenter of socially conservative evangelical Protestantism. In contrast, Massachusetts, Vermont, Connecticut, Oregon, and New Hampshire are relatively liberal and secular. These two sets of states differ along countless dimensions, and among the most prominent is their attitude to sex. Conservative, religious states tend to endorse traditional sexual values while they discourage open and hedonistic attitudes to sexuality, which are far more accepted in liberal, secular states.

One consequence of condemning open sexuality in public is that sexual expression goes underground. For example, teens are more likely to have unprotected sex in conservative states—even when you remove differences in income, education, and access to abortion services from the equation. Religious repression is no match for sex drive—and if anything it seems to exaggerate the urge. This is no surprise to psychologists, who have known for decades that repression doesn't work. It's almost impossible to overcome an addiction by sheer force of will. In 1939, Sigmund Freud first argued that people who rail against an idea are subconsciously drawn to that idea, and two of his disciples, named Seymour Feshbach and Robert Singer, proved him right.

Feshbach and Singer were professors at the University of Pennsylvania in the late 1950s. Experimental ethics laws were lax then, so they devised an unpleasant experiment using electric shocks. One at a time, male psychology students watched a short video of a man completing mental and physical puzzles. A research assistant strapped a small electrode to each student's ankles, which would administer a series of eight shocks as they watched the video. The assistant explained that the shocks would build in intensity, and that it was normal for the students to feel afraid. Half of them were told to express their fears—"to be aware of and admit your feelings." The other half were told to suppress their fears—"to keep your mind off your emotional reactions and not think about them . . . to forget about your feelings . . ." When the video ended, they were asked whether the man they had seen on the video was afraid. As Freud had predicted twenty years earlier, the students who were asked to suppress their own fears believed the man was himself afraid. They were projecting the very emotions they had been asked to suppress onto the world around them. Those who were instead encouraged to express their fears were far less likely to believe the man in the video was afraid. By expressing their own fears, they were freed from the preoccupation with fear that plagued the suppressors.

You might imagine that people in the libertine northeast and northwest states spend more time consuming porn on the Internet, but as Freud predicted long ago, the reverse is true. People from conservative states with traditional views of sexuality are more likely to subscribe to online pornography services. And according to two Canadian psychologists, it's the people from conservative, religious states who search for porn-related terms more often. When Cara MacInnis and Gordon Hodson collected data from Google Trends to examine the search behavior of people from each U.S. state, they found strong correlations between religious belief and porn-related Internet searches, and between conservatism and porn-related searches. In MacInnis and Hodson's own words, "although characterized by an outward and vocal opposition to sexual freedom, regions characterized by stronger political right orientations were relatively associated with a greater underlying attraction to sexual content."

This gap between public and private behavior contradicts the myth that we fail to break addictive habits because we lack willpower. In truth, it's the people who are forced to exercise willpower who fall first. Those who avoid temptation in the first place tend to do much better. That's why heroin-addicted Vietnam vets fared so well when they returned to the U.S. and escaped the drug-taking context altogether, and why it's so important to construct your environment so temptations are far away. According to Wendy Wood, a psychologist at the University of Southern California who studies habits, "Willpower is . . . about looking at those yummy chocolate chip cookies and refusing them. A good habit ensures you're rarely around those chocolate chip cookies in the first place." A combination of abstinence and willpower simply doesn't work. In one study, Xianchi Dai and Ayelet Fishbach at the University of Chicago asked students in Hong Kong to abstain from using Facebook for three days. With each passing day they missed Facebook more acutely, and so inferred that they liked it more, and said they wanted to use it more often. (Students who used other social media sites as substitutes were immune to this effect—but that was only because they found another way to satisfy the same social networking need.)

To understand why abstinence doesn't work, try this simple exercise: For the next thirty seconds, do your best to avoid thinking about chocolate ice cream. Every time your mind's eye conjures the forbidden dessert, wiggle your index finger. If you're like me—and practically everyone else—you'll wiggle your finger at least once or twice. The problem is baked into the task: how can you know whether you're thinking about chocolate ice cream unless you repeatedly compare your thoughts to the one thought you're not allowed to have? You have to think about chocolate ice cream to know whether you were just thinking about chocolate ice cream a second ago. Now substitute chocolate ice cream for shopping, checking your email, checking Facebook, playing a video game, or whatever vice you're trying to suppress, and you'll see the problem.

A psychologist named Dan Wegner first described this puzzle in the late 1980s. The problem, Wegner saw, was that suppression is unfocused. You know what to avoid, but not what to do with your mind instead. When Wegner asked people to ring a bell every time they thought about a forbidden white bear, their bells dinged constantly. But when he told them it might help to think about a red Volkswagen instead, their bells rang half as often. Suppression alone doesn't work—but suppression paired with distraction works pretty well. And what's more, when they were given permission to think about a white bear later, those who had struggled to suppress their thoughts earlier were consumed with the image of the white bear. It was all they could conjure. Meanwhile, the people who were offered a distraction in the form of the red car thought of the white bear occasionally—but they had plenty of other thoughts as well. Suppression isn't just unsuccessful in the short run; as Freud expected, it also backfires in the long run.

T he key to overcoming addictive behaviors, then, is to replace them with something else. That's the logic behind nicotine gum, which serves as a bridge between smoking and quitting. One of the things that smokers miss about cigarettes is the comforting sensation of having the cigarette balanced between their lips—a signal that nicotine will arrive shortly. That sensation continues to give comfort for a while after the smoker quits, which is why you can spot a recent non-smoker by his trail of chewed-on ballpoint pens. Nicotine gum is an effective bridge in part because it administers declining doses of nicotine, but also because it's an oral distraction.

Distraction works just as well if you're trying to overcome a behavioral addiction—if not more so, because you aren't also grappling with substance withdrawal. Take the case of nail-biting. Millions of people bite their nails, and many of those people try a range of remedies that just don't stick. Some paint their nails with a foul-tasting polish, and others swear they'll overcome the habit by willpower alone. The problem with both approaches is that they don't offer a replacement behavior. You might avoid biting your nails because they taste terrible in the short run, but you're really just forcing yourself to suppress the nail-biting urge. We know that suppression doesn't work, so as soon as you stop painting your nails you'll go back to biting them as much if not more than you did before you tried to quit. The urge is so strong in some people that they just go on biting right through the nail polish, forming an oddly positive association between the horrible taste and the relief of satisfying the urge.

A distraction, on the other hand, works quite well. Some people keep a stress ball or a key chain or a small puzzle nearby, so their hands are redirected elsewhere whenever they have the urge to bite. In his book, *The Power of Habit*, the writer Charles Duhigg described this form of habit change as the Golden Rule. According to the Golden Rule, habits consist of three parts: a *cue* (whatever prompts the behavior); a *routine* (the behavior itself); and a *reward* (the payoff that trains our brains to repeat the habit in the future). The best way to overcome a bad habit or an addiction is to keep the cue and the reward consistent while changing the routine—by replacing the original behavior with a distraction. For nail-biters, the cue might be the fidgeting that goes on just before they begin chewing—a subtle search for rough nail-ends that can be smoothed by chewing. Instead of chewing at that point, they might adopt the new routine of playing with a stress ball. And finally, since the reward might be the sense of completeness that comes from chewing the rough nail ends, the nail-biter might complete ten squeezes of the stress ball. So the cue and the reward stay the same, but the routine changes from nail-biting to squeezing the stress ball ten times.

An innovation agency called The Company of Others seems to understand the value of replacing bad routines with good. The agency explains on its website that "we live and think ahead of the trend," and one of those trends is the rise of smartphone addiction. In 2014, The Company of Others launched a product called Realism. Billed as "the smart device for the good of humanity," Realism was designed to treat smartphone addiction. The simple device is an attractive plastic frame that looks like a smartphone without a screen. On one level it's a wry critique of how smartphones remove us from the here and now. Instead of looking at a screen, you could look through a screen-sized frame at what's actually in front of you. And that's how many people respond when they first encounter the device. In a video on the product site, one man says, "Smart devices do get in the way of my relationships with my wife, children, and friends." A woman says, "We don't need to Instagram about our dessert. Nobody cares about our cheesecake."

On a deeper level, though, Realism is to smartphone addicts what nicotine gum is to smokers, and what stress balls are to nail-biters. It's a fitting replacement for a genuine smartphone, because it's roughly the same size, it fits in your pocket, and it gives you many of the same physical feedback cues that come from holding and using a smartphone. What makes Realism appealing is that it obeys the Golden Rule: the cue that leads you to pull out your phone prompts you to pull out the plastic frame instead, which gives you many of the same physical reward cues since it looks and feels a lot like a phone. The cue and reward are intact, but the routine of losing yourself in your smartphone is replaced by a better alternative.

Though the Golden Rule is a useful guide, different addictions demand different routine overrides. What works for people who can't stop checking their emails over lunch may not work for WoW addicts. The key is to work out what made the original addiction rewarding. Sometimes the same addictive behavior can be driven by very different needs. When Isaac Vaisberg reflected on his WoW addiction, he saw that interacting with other players soothed his loneliness. So Vaisberg overcame his addiction, in the long run, by cultivating a vibrant social life and taking on a new job that brought him meaningful extended relationships. Vaisberg was an impressive athlete, so he wasn't particularly attracted to the "crush your enemies" aspect of WoW.

Other WoW addicts, particularly gamers from poorer or working-class backgrounds, are attracted to the element of fantasy that allows them to "travel" to new places they might never otherwise see. Still others are bullied at school, so for them the addiction fills the need for revenge or for physical domination. (Many of these motives aren't psychologically healthy; there's also value in seeing a therapist to address their underlying causes.) Each underlying motive implies a different solution. Once you understand why each addict plays for hours on end, you can suggest a new routine that satisfies his underlying motive. The bullied gamer might benefit from martial arts classes; the frustrated traveler from reading exotic books and watching documentaries; and the lonely gamer from cultivating new social ties. Even if the solution doesn't come easily, the first step is understanding why the addiction was rewarding in the first place, and which psychological needs it was frustrating in the process.

———

Building a new habit is difficult. We know this because the same people seem to make the same resolutions every January. According to one study, roughly half of all Americans make New Year's resolutions—most of them to lose weight, exercise more often, or stop smoking. Three quarters stick to their resolutions through January, but by June roughly half report failing. By the following December most are back making the same resolution they made a year earlier.

One major challenge is that a habit doesn't become routine for weeks or even months. During that fragile early period, you have to be vigilant to protect whatever gains you've made. This is tricky because habit formation takes longer for some people than for others. There isn't a magic number. Several years ago, four English psychologists tracked habit formation in the real world. They asked a group of university students to spend twelve weeks pursuing a new habit in exchange for £30. At the first meeting, each student chose a new healthy eating, drinking, or exercise behavior that might follow a daily cue. For example, some chose to eat an apple with lunch; others running fifteen minutes in the hour before dinner. The students carried out the same behavior every day for eighty-four days, and logged in daily to report whether and how automatically they completed the action.

On average, the students formed habits after sixty-six days. But that average hides how much that number varied. One student took just eighteen days to cement his habit, while the authors calculated that another would need 254 days. Few of the habits were very demanding, and they weren't designed to override existing bad habits, so these numbers are lower than they might be among addicts who are trying to shed chronic addictions. Even if sixty-six days is a reasonable estimate, that's still a long time to maintain a new habit in place of an entrenched, deeply rewarding behavior.

There is one subtle psychological lever that seems to hasten habit formation: the language you use to describe your behavior. Suppose you were trying to avoid using Facebook. Each time you're tempted, you can either tell yourself "I can't use Facebook," or you can tell yourself "I don't use Facebook." They sound similar, and the difference may seem trivial, but it isn't. "I can't" wrests control from you and gives it to an unnamed outside agent. It's disempowering. You're the child in an invisible relationship, forced not to do something you'd like to do, and, like children, many people are drawn to whatever they're not allowed to do. In contrast, "I don't" is an empowering declaration that this isn't something you do. It gives the power to you and signals that you're a particular kind of person—the kind of person who, on principle, doesn't use Facebook.

We know this works because two consumer behavior researchers, Vanessa Patrick and Henrik Hagtvedt, ran an experiment using the technique. They asked a group of women to think of a meaningful long-term health goal, like exercising three times a week or eating healthier food. The researchers explained that the women would face challenges on their quest to live healthier lives, and that they should deal with temptation with self-talk.

Faced with the prospect of exercising after a long day of work, for example, one group was told to say, "I can't miss my workouts," while the other was told to say, "I don't miss my workouts." After ten days the women returned to the lab and reported on their progress. Just 10 percent of the women persisted with their goal when they were told to say "I can't," whereas a full 80 percent persisted when they said, "I don't." Their language empowered them rather than implying they were in the grip of an external force beyond their control. This study tracked behavior across just ten days, so it's difficult to draw strong conclusions. The right words seem to help, but overturning addiction is certainly more complicated than saying "I don't" whenever you're tempted to regress.

Even when helpful new habits override harmful old ones, there's a chance they'll become just as addictive. That was the case for Civil War veteran Robert Pemberton, who tried and failed to treat his morphine addiction with cocaine. The goal, in the long run, is to be free of habits altogether—not to replace one habit with another. For all of distraction's benefits, it's a short-term solution that rarely eliminates addiction on its own. The missing piece in the treatment puzzle is to redesign your environment so temptations are as close to absent as possible. That's the idea behind the technique of *behavioral architecture*.

How far are you from your phone right now? Can you reach it without moving your feet? And, when you sleep, can you reach your phone from your bed? If you're like many people, this is the first time you've considered those questions, and your answer to one or both will be "yes." Your phone's location may seem trivial—the sort of thing you'd never bother to consider in the midst of your busy life—but it's a vivid illustration of the power of behavioral architecture. Like an architect who designs a building, you consciously or unconsciously design the space that surrounds you. If your phone is nearby, you're far more likely to reach for it throughout the day. Worse, you're also more likely to disrupt your sleep if you keep your phone by your bed. Nobody knows this better than reSTART's Cosette Rae, whose preference for the clunky 1990s game Myst I mentioned earlier in the book. "I 'purposely' lose my phone during the day," Rae told me when I visited reSTART. "I have to have a smartphone for work, but I refuse to turn on the ringer." I struggled to reach Rae for months before I finally caught her on her office phone at reSTART. She apologized and told me that was the only way she could manage her smartphone addiction.

Behavioral architecture acknowledges that you can't escape temptation completely. You can't stop using your phone altogether, but you can aim to use it less often. You can't avoid checking email, but life should be compartmentalized so refreshing your email account isn't always an option. There's a time for work and tech, and another for unencumbered vacations and social interactions. Many of the tools that drive our addictions are deeply invasive, so we're forced to be vigilant. Smartphones are ubiquitous; if you own wearable tech, it doesn't leave your body while you're awake (and sometimes while you're asleep as well). Work comes home with you in the form of smartphones, tablets, and laptops, and shopping is always an option. It's tempting to sleep with your smartphone nearby "just in case," and recent studies have shown that merely looking at an illuminated screen shortly before bed severely hampers your ability to sleep deeply. These devices are engineered to remain with us at all times—that's one of their key selling points—so it's easy to allow them to pierce the boundaries between the tech-on and tech-off components of our lives.

The first principle of behavioral architecture, then, is very simple: whatever's nearby will have a bigger impact on your mental life than whatever is farther away. Surround yourself with temptation and you'll be tempted; remove temptation from arm's reach and you'll find hidden reserves of willpower. Proximity is so powerful that it even drives which strangers you'll befriend.

When World War II ended, universities struggled to cope with record enrollments. Like many universities, the Massachusetts Institute of Technology built a series of new housing developments for returning servicemen and their young families. One of those developments was named Westgate West. The buildings doubled as the research lab for three of the greatest social scientists of the twentieth century and would come to reframe the way we think about behavioral architecture.

In the late 1940s, psychologists Leon Festinger, Stanley Schachter, and sociologist Kurt Back began to wonder how friendships form. Why do some strangers build lasting friendships, while others struggle to get past basic platitudes? Some experts, including Sigmund Freud, explained that friendship formation could be traced to infancy, where children acquired the values, beliefs, and attitudes that would bind or separate them later in life. But Festinger, Schachter, and Back pursued a different theory.

The researchers believed that physical space was the key to

friendship formation; that "friendships are likely to develop on
the basis of brief and passive contacts made going to and from
home or walking about the neighborhood." In their view, it
wasn't so much that people with similar attitudes became friends,
but rather that people who passed each other during the day
tended to become friends and so came to adopt similar attitudes
over time.

Festinger and his colleagues approached the students some
months after they had moved into Westgate West, and asked them
to list their three closest friends. The results were fascinating—
and they had very little to do with values, beliefs, and attitudes.

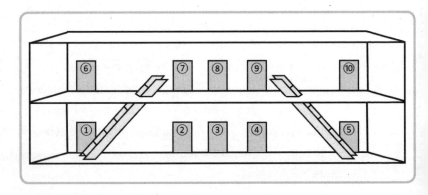

Forty-two percent of the responses were direct neighbors, so the
resident of apartment 7 was quite likely to list the residents of
apartments 6 and 8 as friends—and less likely to list the residents
of apartments 9 and 10. Even more striking, the lucky residents
of apartments 1 and 5 turned out to be the most popular, not
because they happened to be kinder or more interesting, but be-

cause they happened to live at the bottom of the staircase that their upstairs neighbors were forced to use to reach the building's second floor. Some of these accidental interactions fizzled, of course, but in contrast to the isolated residents of apartments 2, 3, and 4, those in apartments 1 and 5 had a better chance of meeting one or two kindred spirits.

Just as we tend to befriend strangers who are nearby, we're also drawn to whatever temptation happens to be within arm's reach. Many remedies for behavioral addiction involve creating psychological or physical distance between the user and the behavioral trigger. A Dutch design studio called Heldergroen has rigged its office furniture to automatically rise to the ceiling at six o'clock every evening. The desks, tables, and computers are connected to strong steel cables that wind upward through a pulley system driven by a powerful motor. After six, the space becomes a yoga studio or a dance floor—or any other activity that thrives on a blank floor plan. German car manufacturer Daimler has a similar email management policy. The company's one hundred thousand employees can set incoming emails to delete automatically when they're on vacation. A so-called *mail on holiday* assistant automatically emails the sender to explain that the email wasn't delivered, and suggests another Daimler employee who will step in if the email is urgent. Workers come back from their vacations to an inbox that looks exactly as it did when they left several weeks ago.

When you set your emails to auto-delete or your office to disappear, you're acknowledging that you're a different person when you're tempted to check your email or work late. You may be an adult now, but this future version of you is more like a child. The best way to wrest control from your childish future self is to act while you're still an adult—to design a world that coaxes, cajoles, or even compels your future-self to do the right thing. An alarm clock called SnūzNLūz illustrates this idea beautifully. SnūzNLūz is wirelessly connected to your bank account. Every time you hit the snooze button, it automatically deducts a preset sum and donates it to a charity you abhor. Support the Democratic Party? Hit snooze and you'll donate ten dollars to the G.O.P. Support the Republican Party, and you'll donate to the Democratic Party. These donations are your present self's way of keeping your future self in line.

SnūzNLūz shapes your behavior with small punishments, promising pain if you misbehave rather than pleasure if you do the right thing. That's a wise choice. Rewards are a lot more fun than punishments, but if you're looking to change a habit small punishments or inconveniences are often more effective. This is an old idea that pervades psychological science: that we're far more sensitive to losses and negative events than we are to wins and positive events. To give you a sense of how this works, suppose you're on a game show and the host offers you the chance to play a game. He produces a coin and tells you that he'll pay you $10,000 if the coin comes up heads, but you'll have to pay him $10,000 if the coin comes up tails. Would you play the game? Very few people say yes even though the game is fair—much fairer than most casino games, which are rigged in the house's favor. But the prospect of losing $10,000 is much more daunting than the prospect of winning $10,000 is appealing. Your mind will gravitate toward the loss, focusing on the potential pain of losing far more than on the potential joy of winning. Losses are all-encompassing, and we'll do a lot to avoid them. (I've asked hundreds of people whether they'd play this theoretical game, and only 1–2 percent say yes. To get half of the room to play, the potential win has to be about two and a half times greater than the potential loss.)

Maneesh Sethi is an entrepreneur who designed a product called Pavlok, which uses the power of negative feedback to discourage bad addictive habits. "There are two kinds of people," Sethi told me. "People who generate lots of ideas, and people who can execute those ideas." Sethi describes himself as an ideas man. "A few years ago, I hired a girl to slap me in the face every time I went on Facebook." That worked well, for a while, but Sethi developed a more permanent solution in the form of Pavlok, a small wearable wristband that gives feedback whenever the wearer engages in a forbidden bad habit. This is known as aversion therapy: pairing an action you'd like to change with an unpleasant or aversive sensation. At the subtle end of the spectrum, Pavlok beeps or vibrates when you do something you've pledged not to do, and at the invasive end it delivers a moderate electric shock, or zap. Users can administer the negative feedback manually, or they can pair the device with an app that automatically delivers the feedback in response to pre-determined cues.

Sethi generously sent a Pavlok after we finished speaking (the device retails for five hundred dollars). I tested the zap function as soon as I opened the box. It was surprisingly strong, and I understood how a regular dose of zaps might discourage bad habits. When Richard Branson tried the device, he reportedly punched Sethi in the stomach in surprise at the shock's potency. Other users include entrepreneur and writer Tim Ferriss, actor

Ken Jeong, businessman Daymond John, and Congressman Joe
Kennedy.

Pavlok has shown early promise, but it's too early to say
whether it will achieve mainstream appeal. (One psychiatrist in
New York City has begun using the product, but it's still considered experimental.) Just as the first iPad generation hasn't come
of age yet, the remedies designed to dull behavioral addictions
are still immature. All of these proposed solutions are to some
extent still exploratory, and Sethi and his team are always tweaking Pavlok and its app. Still, the product's funding campaign on
Indiegogo was a huge success, raising almost $300,000—more
than five times the amount Sethi sought when he launched the
campaign.

Part of the reason for Pavlok's success are the product's simplicity and the vivid testimonials on the product's site. Here's how
Sethi describes the product on its webpage:

Here's how it works:

1. Download the app, and choose the habit you want to
   break.

2. Wear your Pavlok and listen to the 5-minute audio training sessions. The app will automatically trigger the Pavlok, you simply need to pay attention.

3. Use Pavlok's zap when you do the bad habit. Pavlok can
   be triggered by sensors and apps, a remote control, and
   manually. Manual is as effective as automatic.

4. It might seem like your habit is broken in 3-4 days. Continue doing the bad habit (with zap) for at least five days, making yourself do the bad habit on purpose if necessary. The longer you continue, the more permanently the habit stays broken.

Sethi says that early results are promising. Only a few percent of smokers manage to quit cold turkey, but Sethi reports that 55 percent of a sample of regular smokers quit smoking after following the Pavlok's five-day training process. The same is true for other behaviors. In videos on the app, Nagina explains how she stopped biting her nails, David stopped grinding his teeth, and Tasha explains how she stopped eating sugar. Writing for Yahoo Tech, Becky Worley explained that Pavlok's shocks discouraged her from using Facebook, which she felt she was using far too often. It's too early to know whether Pavlok will work for everyone as well as it did for Becky, Nagina, David, and Tasha, but the science behind the device is sound. Even without the Pavlok itself, you can engineer your environment so you follow bad habits with mild punishments—tasks you'd prefer to avoid or experiences that you find unpleasant.

One of Pavlok's biggest strengths is that it does the hard work for you. You don't have to remember to do the right thing, because the device will remind you with a zap when you fail. But it also has a weakness: you can stop using it whenever you like. Punishments are effective when they're genuinely unpleasant, but some people might stop using a device that makes them feel bad. The trick for those people is to find a method that isn't aversive.

I was wrapping up my PhD at Princeton University in 2008,

when Nobel prizewinner Daniel Kahneman invited me to his office. "You can tell me about your research," he said. I was excited. Kahneman and his colleague Amos Tversky had pioneered the field of judgment and decision making, and now, forty years later, I was a young researcher in the same field. I told Kahneman that I wanted to invent a tiny alarm clock that followed each of us around and rang whenever we were about to make an important decision. He and Tversky had spent decades studying laziness in decision making, so he understood what I was getting at. "So the alarm clock will tell people when to pay close attention?" he asked. "You need the mental equivalent of a sign that flashes the words PAY ATTENTION NOW! in front of people's eyes at exactly the right moment."

I still haven't invented the alarm clock, but a company called MOTI is piloting a device (also called MOTI) that comes close. The company's founder, Kayla Matheus, noticed that people tended to abandon wearable tech over time. "When you look at the research in wearables," she said in an interview with Fast-CoExist, "there's a huge drop-off rate. Data alone isn't enough. We're human beings—we need more than that." Matheus was speaking from experience. She'd torn her ACL and was struggling to keep up with her rehab. Many people have the same experience with fitness trackers, which they buy and then quickly abandon at the bottom of a drawer. Fitness trackers are passive devices: you have to choose to use them, otherwise they're useless.

Matheus designed MOTI to reinforce good habits in the same way that Kahneman's PAY ATTENTION NOW! reinforces careful thought. It's a simple animal-like gadget that tracks behavior over time. "It will basically learn what's normal for you," says Matheus. "If you start straying, then you're going to get prompted with a reminder. Rather than being a push notification you can easily wipe away, MOTI might get sad or angry." There's a small button on the front of the device, which you push when you've done the right thing. For some people, that's rehab exercise for a torn ACL; for others, it's running once a day, or shutting down their smartphones and laptops and going to bed before 10 P.M. MOTI flashes a rainbow of colors and emits a series of happy chirps when you do the right thing; when you leave it unattended for a while, it flashes red and chirps and buzzes less happily as a reminder. Unlike passive apps, MOTI sits on display. You can't ignore it—and early testing suggests that people form a bond with the device, so they don't abandon it. One of Matheus's early testers was struggling to drink enough water. "He tends to get stuck at his desk and forgets to hydrate," Matheus says. "Because it's a physical object, all of a sudden it becomes an environmental cue. Whenever he's typing at his computer, his eyes happen to flick over to MOTI, who's right there, and he'll be reminded." MOTI's testers seem to feel a sense of duty to the little device—as though they're disappointing it when they do the wrong thing.

In fact, choosing rewards and punishments that also affect someone else you care about is a very effective way of forming the right habits. That's the idea behind a technique called the Don't Waste Your Money motivator. To begin, you set a goal. Say you've been using your smartphone for an average of three hours a day, and over the next four weeks you'd like to lower that number by fifteen minutes each week. By the end of the four-week period, you'd like to be using your phone for an average of no more than two hours a day. Each week, you put a sum of money into an envelope—say, fifty dollars. The sum should feel significant, but not so large that losing it four weeks in a row is financially crippling. You stamp the envelope, and address it to a frivolous organization, or a cause that you don't support. (Think the SnūzNLūz, where Republicans donate to the Democratic Party, and vice versa.) Among other organizations, one manual suggests the following:

> American Yo-Yo Association
> 12106 Fruitwood Drive
> Riverview, FL 33569

> Fabio International Fan Club
> Donamamie E. White, President
> 37844 Mosswood Drive
> Fremont, CA 94536

If, on the other hand, you achieve your daily usage goal, you tear open the envelope and spend the money *relationally*: you take a friend to lunch, buy your son an ice cream, or buy your spouse a gift. Relational spending has two advantages: it makes you accountable, so failing to reach your goal also hurts someone else; and it's a superior reward, because spending on others makes you happier than spending on yourself or paying your bills.

Behavioral architecture acknowledges that you can't avoid temptation completely. Instead of abstinence and avoidance, many solutions come in the form of tools designed to blunt the psychological immediacy of addictive experiences. Benjamin Grosser, a web developer, devised one of these clever tools. Grosser explains on his website:

> The Facebook interface is filled with numbers. These numbers, or metrics, measure and present our social value and activity, enumerating friends, likes, comments, and more. Facebook Demetricator is a web browser add-on that hides these metrics. No longer is the focus on how many friends you have or on how much they like your status, but on who they are and what they said. Friend counts disappear. "16 people like this" becomes "people like this." Through changes like these, Demetricator invites Facebook's users to try the system without the numbers, to see how their experience is changed by their absence. With this work I aim to disrupt the prescribed sociality these metrics produce, enabling a network society that isn't dependent on quantification.

The Demetricator makes it impossible to check how many likes or comments or friends you have. Here's one screenshot using Facebook's regular metrics:

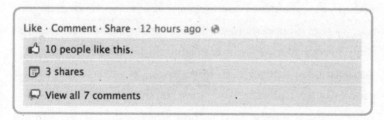

Like · Comment · Share · 12 hours ago · 🌐
👍 10 people like this.
📄 3 shares
💬 View all 7 comments

Everything is measured numerically, and it refreshes as time passes. There's always something to check, because the feedback

changes with each new like or comment. In contrast, here's the
same feedback filtered through Grosser's Demetricator:

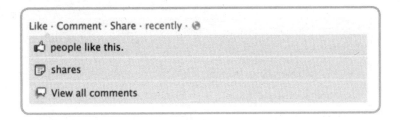

You learn that people like your post, that it's been shared, and
that there are comments, but you can't fixate on the numbers because they've vanished. The Demetricator does exactly the opposite of what Fitbits and Apples Watches do. When we buy those
devices we choose to inject new metrics into our lives—to measure how far we walk, how deeply we sleep, how fast our heart
beats, and so on. These are processes that, for millennia, went
unmeasured and untracked.

Grosser's Demetricator is relatively subtle. It dulls the feedback cues that make Facebook addictive, rather than preventing
you from using Facebook altogether. If demetrication isn't potent
enough, the WasteNoTime program is a heavy-handed alternative. WasteNoTime monitors how long you spend on sites that
you add to a Block List. You might add Facebook, Twitter, and
YouTube to your Block List, for example. You can outright block
your browser from accessing some of those programs, and for
others you can impose a usage limit. For example, you might use
the rule, "between 9 A.M. and 5 P.M., I will spend no longer than

thirty minutes on Facebook." You can set stringent limits during work hours and before bed, and lenient limits during leisure time. There are ways around WasteNoTime in an emergency, but evading the program is frustrating enough that it serves as a strong deterrent.

Smart behavioral architects do two things: they design temptation-free environments, and they understand how to blunt unavoidable temptations. This process is a bit like taking apart a computer: by reverse engineering the experience, you learn what makes it addictive in the first place, and therefore how to defuse it. Take the case of binge-viewing on Netflix. You may not want to avoid watching Netflix altogether, so how do you fight the tide of episode-ending cliffhangers? If you understand the structure of binge-viewing, it becomes easier to avoid falling into the binge-viewing trap. This is the basic structure of two episodes within a series (and the beginning of a third):

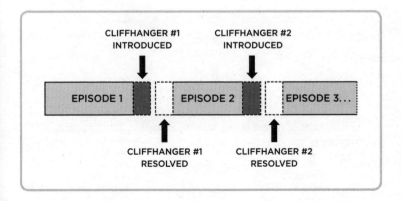

An episode typically lasts forty-two minutes (with roughly eighteen minutes occupied by ads). The last few minutes of the first episode are devoted to setting up and introducing the first cliffhanger: someone's shot and we wonder whether he's still alive, or the killer is unmasked but we can't see his identity. Then, the first few minutes of the second episode are devoted to resolving that first cliffhanger, so the viewer can move on to the meat of the second episode in anticipation of—you guessed it—the second cliffhanger, which arrives near the second episode's end. For a viewer, this is catnip. Assuming you enjoy the show, if you obey the structure of the episodes laid out by the writers, you'll struggle to escape the binge-viewing process. What you can do instead, though, is to short-circuit the cliffhangers either before they're introduced or after they're resolved. There are two ways to do this. Instead of watching each forty-two-minute episode from start to finish, you can watch the first thirty-seven minutes of each episode, turning off the show before the cliffhanger arrives. (If you know what to look for, you'll often see the cliffhanger sneaking up on you.)

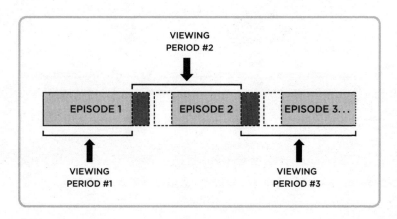

Or, if you aren't sure you'll be able to stop before the cliff-hanger arrives, you can watch the beginning of the next episode and stop after the cliffhanger has been resolved. That way you watch from the fifth minute of every episode to the fifth minute of the next one. This approach doesn't diminish the pleasure of viewing—you still get to enjoy the cliffhanger and its resolution—but it does limit your chances of bingeing.

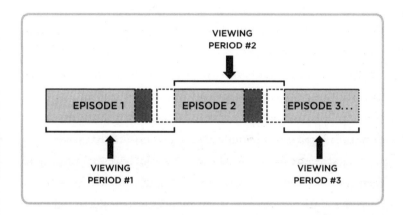

The problem for most of us most of the time is that these experiences are so new to us that we aren't sure where to begin. But once you understand how cliffhangers—or any other addictive devices—work, you can find ways around them. Sometimes it's best to watch the experts. When Bennett Foddy told me he decided not to play World of Warcraft, he was making a difficult and careful decision. One classic test for deciding whether to take up a new game or activity is to ask yourself whether you could afford to lose a certain amount of time to the experience *today*. According to a phenomenon known as the planning fallacy, even if we're short on time today, we assume we'll have more time in the future. That's why people say "no" to many requests that fall within the next week, but "yes" to similar requests that fall several months in the future. This is a mistake, because your level of spare time today is an excellent guide to how much time you'll have in a couple of months. If you're concerned that WoW might suck up too much time today and tomorrow, you should have the same concern about how it might affect you in two months or a year or two years. That's why Foddy was right to avoid WoW, and why saying "no" to a potentially addictive time-hungry experience is a wise move.

One of the problems with WoW is that it swamps your schedule. You have to play when your friends are playing, so more pressing tasks fall by the wayside. In contrast, the advent of on-demand viewing and digital video recorders (DVRs) means you can delay watching a TV show until you have nothing more pressing to do. DVRs seem like godsends, but they're actually powerful addiction drivers. TV networks historically saved their biggest and best shows for coveted prime-time slots, and though people recorded shows on their VCRs, the process was far more cumbersome than DVR recording and on-demand viewing are now. Today, you can find a string of major shows between two and six in the morning, when viewership declines to its lowest levels. *Mad Men*, one of the biggest shows of the past decade, began showing older seasons in the middle of the night to allow new viewers to catch up to the current season. The result was that thousands of people who once might have missed the boat can decide whether they want to invest in watching the show from scratch. Many of these people are classic binge-viewers who wait till TV shows are vetted by first movers before they decide to watch as well. Instead of missing the show altogether—and spending that time doing other things—they're roped in and forced to binge-watch if they want to catch up to the show's current episodes. The solution here is not to swear off using the DVR, perhaps, but to use it sparingly and mindfully. Or to use Bennett Foddy's test: if it's going to take too much time now, it's not wise to record it for a week or a month in the future.

It's also easy to tell yourself you'll watch just an episode or two, and unless the show is really worth your time you'll resist watching the remaining episodes. In a recent study, though, Netflix measured how long it took its viewers to become hooked to each show. For each show, Netflix calculated how many episodes it took for 70 percent of its viewers to continue on to the end of the first season or beyond. Most shows weren't quite addictive after just the pilot episode, but some became addictive to a great majority of viewers by episodes two, three, or four (see figure on page 292).

This leaves you with three options: avoid watching the show altogether, begin watching when you can afford to lose several hours to a binge session, or—best of all—use the cliffhanger-disarming technique to defang end-of-episode cliffhangers. If you design your environment wisely, you'll stand a better chance of avoiding harmful behavioral addictions.

But not all addictive experiences are bad. In theory, the same hooks that drive addiction can also be harnessed to drive healthier eating, regular exercise, retirement saving, charitable giving, and committed studying. Sometimes, the problem isn't that we're

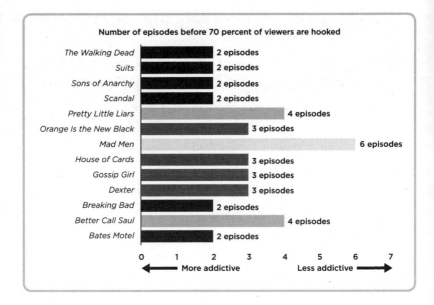

addicted to the wrong kinds of behaviors, but rather that we abandon the right kinds. Behavioral architecture isn't just a tool for doing less of the wrong things; it's also a tool for doing more of the right things.

Enter gamification.

12.

Gamification

In late 2009, Swedish ad agency DDB Stockholm launched an online campaign for Volkswagen. Volkswagen was releasing a new eco-friendly car that was designed to make driving more fun, so DDB named the campaign The Fun Theory. "Fun can change people's behavior for the better," one executive explained, so perhaps a dose of fun would nudge drivers to try the new car. To generate buzz, DDB launched a series of clever experiments around Stockholm. Each one turned an otherwise mundane behavior into a game.

The first experiment took place at central Stockholm's Odenplan metro station. Commuters had two options when exiting the station: to walk up a bank of twenty-four stairs, or to stand still on a narrow escalator. Surveillance footage showed that commuters were lazy by default, piling onto the crowded escalator rather than taking the empty staircase. The problem, DDB explained, is

that stairs aren't fun. So, late one evening, a team of workers converted the staircase into an electronic piano. Each stair became a
piano key that played a loud tone in response to pressure. In the
morning, commuters approached Odenplan's exit as they usually
did. At first, most took the escalator, but a few happened to take
the stairs, unintentionally composing brief melodies as they left
the station. Other commuters took note, and soon the stairs were
more popular than the escalator. According to the video, "66 percent more people than normal chose the stairs over the escalator."
People flock when you turn a mundane experience into a game.

DDB released other experiments as the campaign gathered
steam. At a popular park, an electronics expert created the "deepest bin in the world"—a trash can rigged to emit an echo implying that each piece of garbage plummeted before crashing far
below. Other cans in the park attracted eighty pounds of trash
each day; the deepest can attracted twice as much. Elsewhere,
people were misusing recycling bins around the city, so DDB
turned one bin into an arcade game. The game rewarded people
who used the bin correctly with flashing lights and points that
were recorded on a large, red display. An average of just two people used most nearby bins correctly each day; more than a hundred people used the arcade bin correctly each day.

The campaign was wildly successful. The videos attracted a
combined total of more than thirty million YouTube hits, and
plenty of online buzz. In 2010, DDB won the Cyber Grand Prix
Lion at the world's largest advertising festival—an enormous
honor bestowed on the "world's most celebrated viral campaigns."
Beyond industry plaudits, the campaign also changed how peo-

ple behaved. For a brief time, the people of Stockholm were ever so slightly greener and healthier.

———

There are two ways to approach behavioral addictions: eliminate them or harness them. Elimination was the subject of the first eleven chapters of *Irresistible*, but—just as DDB did in Stockholm—it's possible to channel the forces that drive harmful behavioral addiction for the good. The human tendencies that enslave us to smartphones, tablets, and video games also prepare us to do good: to eat better, exercise more, work smarter, behave more generously, and save more money. To be sure, there's a fine line between behavioral addictions and helpful habits, and it's important to keep that line in mind. The same Fitbit that fuels exercise addiction and eating disorders in some people pushes others to leave the couch behind during an hour of exercise. Addictive levers work by boosting motivation, so if your motivation is already high there's a good chance those levers will compromise your well-being. If you're a couch potato who hates to exercise, a dose of motivation can only help.

A broad survey of human behavior reveals plenty of room for improvement. Sixty percent of the world's developed population is overweight or obese, including 67 percent of Americans, 66 percent of New Zealanders, 65 percent of Norwegians, and 61 percent of Brits, Germans, and Australians. Graduation rates in the United States are declining at every education level, from elementary school to four-year colleges. In response, the National Center for Public Policy and Higher Education expects personal income to fall over the next fifteen years. Americans save just 3 percent of their household income; Danes, Spaniards, Finns, Japanese, and Italians save even less. A paper in the prestigious medical journal the *Lancet* predicted that half of all babies born in the world's developed countries after 2000 will live past one hundred years, which will outstrip their retirement savings by decades. Between 2013 and 2015, U.S. residents were among the one or two most generous nations in the world—but even so, Americans donated less than 2 percent of their income to charitable causes.

Almost everyone wants to change at least one behavior. For some, it's spending too much and saving too little; for others, it's wasting nine tenths of the work day checking emails; for others, still, eating too much or exercising too little. The obvious path to change is with effort, but willpower is limited. People are more likely to do the right thing, DDB showed, if the right thing happens to be fun. A computer programmer named John Breen had the same intuition when his son was struggling to learn SAT vocabulary in 2007. Breen designed a computer program that presented his son with randomly chosen words, and asked him to choose the best definition for each word among four alternatives. Breen also ran a website that educated people on world poverty, so he decided to combine the two. If the site attracted enough traffic, he could sell banner ad space to the highest bidder, and use the ad revenue to buy rice for the needy. And so FreeRice.com was born.

For every correct answer, Breen promised to donate ten grains of rice to a food charity. The site launched on October 7, 2007, and on its first day raised 830 grains of rice. FreeRice grew so fast that two months later Breen raised three hundred million grains in a single day. In 2009, he offered the platform to the United Nations World Food Programme, and in 2014 the site raised its hundred *billionth* grain—enough to feed five million adults for a day.

When American students are forced to learn thousands of words for the SATs, that's a chore; yet that's exactly what thousands of FreeRice users choose to do with their free time every day. The site succeeds because Breen managed to turn the chore into a game. All the elements are there: each correct answer earns ten points (depicted as grains of rice), which function like a game score. You can track how many correct answers you've delivered in a row, and the game reports your longest winning streak. Meanwhile, the words become more obscure as you rise through the game's sixty levels—and decline in obscurity when you make a mistake. That way the game is always pitched perfectly between too easy and too difficult. Breen wisely added graphics as well, so you can track your progress visually: a small wooden bowl fills with rice until you reach one hundred grains, and then places the ball of one hundred grains beside the bowl as it begins filling again. Reach a thousand grains and an even bigger pile forms next to the bowl. Some users form groups that play together—the highest-scoring groups and individuals appear on a daily leaderboard—and you and your group members can stop and start whenever you like. FreeRice looks like a combination of learning and giving, but under the hood it's driven by a gaming engine.

What DDB did for Volkswagen and Breen did for FreeRice is known as *gamification*: taking a non-game experience and turning it into a game. A computer programmer named Nick Pelling coined the term in 2002. Pelling realized that game mechanics could make any experience more compelling, but he struggled to commercialize the concept, which lay dormant until Google and several prominent venture capitalists revived it in 2010. The central theme of gamification is that the experience itself should be its own reward. Even if you aren't motivated to donate to a food charity, or to learn new words, you should want to spend your time playing FreeRice. Over time, despite yourself, you'll find that you *are* learning and donating rice.

Gamification researchers Kevin Werbach and Dan Hunter examined over one hundred examples of gamification, and identified three common elements: points, badges, and leaderboards. PBL, as the triad is known, first came together in airline frequent-flyer programs. United launched the first airline loyalty program in 1972, long before the advent of gamification, and other airlines soon introduced similar programs. With each flight or qualifying purchase, flyers earn *points* in the form of miles; when they earn enough points in a single year, they win *badges* in the form of status markers—silver, gold, platinum, and so on; and high-status members stand in different lines, board the plane first, and sometimes receive special treatment on the airplane—rewards that function as a conspicuous *leaderboard*.

Gamification is a powerful business tool, and harnessed appropriately it also drives happier, healthier, and wiser behavior.

That philosophy drove Richard Talens and Brian Wang, who met in 2004 as freshmen at the University of Pennsylvania. Talens and Wang had two things in common: they both loved video games, and they were both fitness fanatics. "We kind of recognized each other because we saw each other eating broccoli and tuna in the cafeteria," Talens recalled in an interview. "We had a very similar mentality to fitness because we both grew up very out of shape. We both grew up playing video games, and we both saw fitness as a game." Talens and Wang became amateur body builders, and in 2011 they launched a gamified fitness website called Fitocracy. By 2013, Fitocracy had one million users; by 2015, two million.

Fitocracy rewards users with points after every workout— more points for harder workouts—and badges when those workouts reach certain milestones. Run a 10k, for example, and the site gives you the 5k badge, the 10k badge, and awards you 1313 points. When you go to the gym you'll see two kinds of people: people who prefer to work out alone; and people who turn fitness into a social event. Fitocracy appeals to both kinds of people by giving you the opportunity to interact with other users. You can challenge them to duels and discuss your latest workout—or you can treat the site like a private activity log, challenging yourself to run farther and lift more without having to share your progress with anyone else. Variety is also a critical gamification ingredient, and Fitocracy injects variety by allowing you to adopt quests and challenges that draw on your favorite exercises. Wang and Talens have collected dozens of stories of people losing a hundred pounds with the site's help—the majority having struggled to stick to an exercise regime for years.

Many adults fold in the face of temptation, so you can imagine how children struggle to do the right thing. Adults make wise decisions at least some of the time because they're able to look into the distant future. Children, on the other hand, make decisions that suit them right now. There is no long view where children are concerned, so a chocolate cake is all temptation and no downside. But children love games as much as adults do, so gamification endows children with a dose of self-control. Take the case of dental hygiene. Kids have better things to do than brushing their teeth, particularly just before bedtime. Enter Philips Sonicare, which released a gamified toothbrush in August 2015. The toothbrush is designed to encourage kids to brush for a full two minutes. It has a small screen that displays a character called a Sparkly. Kids earn points for brushing each quadrant of teeth, and those points feed the Sparkly. The Sparkly proved so endearing that kids couldn't get enough of the toothbrush. In an interview, a veteran at the company said, "Because kids love the game and they interacted so much with the app they didn't go to bed right away." The app had to be altered so the updated Sparkly falls over with exhaustion after brushing ends.

As NYU Game Center director Frank Lantz told me, designing games is tough. For every game that hooks the masses, thousands go largely unplayed. Philips had the opposite problem, deliberately tweaking the Sparkly app to make it *less* addictive. These tweaks are a common feature of gamified platforms, because it's difficult to predict which elements will drive behavior. In 2009, Adam Bosworth, the former head of Google Health, launched a health app called Keas. At first, Keas was big on data and small on gamification. Bosworth designed the app to deliver mountains of feedback tailored to each user. Users completed quizzes and entered their workouts and meals, and Keas explained how their choices shaped important health outcomes. In Bosworth's mind, users would exercise more and eat less if they were forced to confront the effects of laziness and gluttony. But idle data reports weren't enough to change behavior, so Keas changed direction. Bosworth rolled out the app at a number of large corporations, where he encouraged employees to form rival teams. Good behavior earned players points, and the new version of Keas incorporated game levels and strategies. Bosworth wanted to make sure that the app had plenty of quizzes, so his team designed many more than he expected users to complete during the app's standard twelve-day program. He undershot the mark: many users devoured the entire set in under a week.

Keas works in part because it's simple. It relies heavily on a four-item quiz that users complete at the beginning and end of the twelve-day program. The questions are:

1. Are you a non-smoker?
2. Do you eat more than five servings of fruit and veggies a day?
3. Do you have a healthy bodyweight (Body Mass Index less than 25)?
4. Do you exercise regularly (more than 45 minutes, 5 times a week)?

For each "yes" answer, users earn a point—so scores of zero or one indicate an unhealthy lifestyle, while three or four indicate healthy behavior. Pfizer, the world's largest pharmaceutical research firm, invested in the app several years ago. Before the program began, 35 percent of its workforce scored zero or one on the app—afterwards, that number fell to 17 percent. Meanwhile, healthy responses (scores of three or four) rose from 40 percent to 68 percent.

Keas operates for profit. Executives at companies like Pfizer pay a fee to use the program, and in turn their employees are healthier, more productive, less likely to call in sick and to draw from the company's healthcare budget. Similar apps work just as well in not-for-profit contexts. An app called Health Lab improves the health of children in low-income communities, and the U.S. government has considered using games to fuel healthy childhood behavior across the country.

In the fall of 2009, a new school opened its doors in New York City. Quest to Learn (Q2L) welcomed seventy-six sixth graders in its first year, and then added one new class at the beginning of each new year. Q2L was the brainchild of several organizations that came together to design a new model of education. The old model, they reasoned, was far from perfect. For centuries, schools had been wrestling with kids who were distracted, unmotivated, and often unhappy in the classroom. School seemed to be unpleasant by design: a combination of rote learning and brute-force instruction. Fun was an afterthought, if anyone considered it at all, so most kids saw school as a chore.

Q2L is different. Like DDB for its Volkswagen campaign, the school was founded on fun. If kids enjoyed school, surely they'd be happier and more engaged. The school's founders decided that the best way to inject fun was to make the learning experience one big game. Learning, it turns out, is ripe for gamification. Each new module of information can be structured like a game that begins at zero knowledge and ends at perfect comprehension. Q2L uses the same gamified structure for each larger learning module, or *mission*: students complete a series of smaller *quests* during the mission period (say, ten weeks), and then finish with a *boss level* that pushes them to apply what they've learned to a new context. The boss level concept draws on classic gaming theory: that players hone their skills by defeating easier opponents before tackling a formidable "boss." The boss serves as a capstone—a signal that the player has completed the mission and can move on to the next one.

In one mission, "Dr. Smallz," sixth or seventh graders learn about the human body. Dr. Smallz has shrunk himself to save an ailing patient, but alas, he has amnesia. The mission lasts thirteen weeks, and across its seven quests, students are charged with several goals: to help Dr. Smallz work out where he is in the patient's body, remind him what each body organ and system does, help him solve the medical mystery of the patient's illness, and, based on what they've learned about the body's anatomy, help him find a way out of the patient's body. By the end of the mission, students have learned the same scientific information that other schools teach, but for them the process is a game. In one task, for example, students build a cell from puzzle pieces. As they research information on each structure within the cell, they earn a piece, and so move closer to completing the task. In another, they learn about the immune system by playing a board game called Virus Attack. The game, designed by the Institute of

Play, asks them to kill a virus by generating white blood cells, antibodies, and T-cells. Students earn rewards and track their progress just as they might do if they were playing a game outside the classroom.

A seventh-grade unit teaches students about the American Revolution. Their mission is to mediate a disagreement among several ghosts at the Natural History Museum. Each ghost represents a different Revolutionary character: a loyalist, a patriot, a landowner, a merchant, and a slave. They disagree about what happened during the Revolution, and the students must collect as much information as possible to prevent the bickering ghosts from destroying the museum's entire collection. The students learn about the American Revolution, but they also learn that the truth is complex; that different parties may view the same event differently, and how to resolve those conflicts.

Q2L's approach seems to be working. The school's math team placed first in the New York City Math Olympiad three years in a row, and its students score roughly 50 percent higher than the average school on New York City's standardized exams. By one metric, students grow intellectually between eighth and tenth grade as much as the average college student grows across all four years of college. Students and their teachers are also engaged: average student attendance sits at an impressive 94 percent, and the school has retained 90 percent of its teachers.

Gamified education sounds like an approach that might appeal most to children, but it works among young adults, too. In 2011, Rochester School of Technology's School of Interactive Games and Media introduced a program called Just Press Play. The program motivates students by introducing voluntary quests.

Each professor introduces these quests, and students have the option to pursue them or to ignore them. Many of the quests are designed for the entire cohort, rather than just for one or two students. For example, the "Undying" quest promises to give all students a reward if 90 percent of the first-year class passes its notoriously difficult introductory course. Historically the pass rate was lower than 90 percent, but the program was so appealing that several juniors and seniors showed up at the freshman computer lab to coach the freshmen through their course. Juniors and seniors weren't eligible to benefit from the quest, but they were so impressed that they were moved to participate. Freshmen graduated at an unprecedented rate that year, and the juniors and seniors offered to help the next year as well. That's the hallmark of a game that works: people are motivated intrinsically, even when they aren't capable of earning extrinsic rewards. My favorite quest is promoted by Professor Andy Phelps himself, one of the program's founders. Phelps's quest is called "A Walk into Mordor," named for a dark and dangerous region in J.R.R. Tolkien's *Lord of the Rings*. "Find my office in the depths of Mordor, when the Black Gate is open," Phelps says. "Get the card. Feel free to strike up a conversation . . ." Students don't even realize they're learning when they meet the professor—as far as they're concerned, they're just completing another quest.

———

From Andy Phelps's quest to Q2L's missions, gamification is designed to raise productivity where people would prefer to be lazy. In many contexts, laziness is the human default. Social psychologists Susan Fiske and Shelley Taylor describe humans as *cognitive misers* to suggest that we avoid thinking the way a miser avoids spending. Indeed, people prefer to think only as much as necessary to reach a just-acceptable conclusion. Miserliness makes sense from an evolutionary perspective, because thinking is costly. It stops an animal from acting, which leaves it vulnerable to predators and less prepared to seize limited opportunities. That's why we rely so heavily on mental shortcuts, stereotypes, and rules of thumb, which allow us to make sense of a complex world as quickly as possible.

This laziness explains why work is dressed like a game. Salaries (*points*) rise with seniority (*levels*), which brings promotions and new titles (*badges*). The difference between most workplaces and genuine games is that people don't go to work because they're intrinsically motivated by the game; instead, the game is how the employer doles out the extrinsic rewards of money, prestige, and praise. As Nick Pelling explained when he coined the term, you'll know you're looking at gamification when the fun of playing the game becomes the reward. In some contexts, gamification can be dangerous. Exercise addicts tend to focus on the game of working out every day, or racking up a certain number of steps or miles. They forget that exercise is primarily designed to make them healthier, developing stress-related injuries instead in the quest for arbitrary fitness goals.

Beyond personal fitness devices, some companies gamify the workplace to motivate their employees. In 2000, four tech entrepreneurs formed a remote call center called LiveOps. LiveOps enlists more than twenty thousand everyday Americans to make telemarketing phone calls, and, more recently, to run the social media platforms of large organizations from Pizza Hut to Electronic Arts. The company vets agents before admitting them to its staff, and once accepted they can work as much or as little as they like in blocks of thirty minutes. All agents need are a landline phone, a computer, a high-speed Internet connection, and a corded headset. Some companies that use LiveOps pay by the minute—for example, twenty-five cents per minute spent on the phone—while others pay per call or per sale. LiveOps appeals to people without a fixed schedule—people who are employed part-time, at home with children, or between steady jobs.

The company's flexibility is a strength, but call center workers without a fixed schedule tend to suffer dips in motivation. To combat those dips, LiveOps introduced a gamified dashboard. Each worker's dashboard contains a progress bar with the percentage of calls that produce sales, trophies and badges for reaching certain sales milestones, and individual challenges tackled and accomplished. A leaderboard broadcasts the top salespeople across the company. According to LiveOps, these game elements improved service ratings by 10 percent, and lowered customer wait times by 15 percent. Sales conversion rates rose, and workers reported feeling more positive about working for the company.

Other organizations have grown by introducing gamified rewards. After Rodney Smith, Jr., noticed a ninety-three-year-old woman struggling to mow her lawn in Huntsville, Alabama, he created an organization known as Raising Men Lawn Care. Raising Men employs young men, many from underprivileged backgrounds, to mow lawns free of charge. (The organization is funded by well-wishers on its GoFundMe page.) The boys are motivated to do the right thing, but they're also motivated by a badge system that borrows from martial arts. As the company's

Facebook page explains, the color-ranking system is "similar to how they do it in martial arts . . . the kids will start the program with a white shirt. Then when they do 10 lawns, they will get an orange shirt, when they do 20 lawns they will get a green shirt, 30 they will get a blue shirt, 40 a red shirt and 50 + lawns will get them a black shirt." Raising Men's success has spawned new chapters across the country, a growing online following, and tens of thousands of dollars in funding.

Gamification doesn't help much when an experience is already fun; it does its best work when the experience is boring. On-the-job training is perhaps the most notoriously boring part of work. At the same time, training is critically important, because poorly trained workers are less productive and less safe. A number of companies are starting to train their employees with games. The Hilton Garden Inn, for example, hired the Virtual Heroes game design studio to develop a virtual training hotel. The game puts team members in a three-dimensional, virtual Hilton Garden Inn hotel, where they serve guests within a timed deadline. Their responses are graded for speed and appropriateness, which translate into Satisfaction and Loyalty Tracking (SALT) scores. The hotels assess employees with SALT scores in the real world, so the virtual game environment is an excellent simulation. Since Hilton's success, Virtual Heroes has taken on a raft of large corporate clients, including the U.S. Army, the Discovery Channel, the Department of Homeland Security, BP, and Genentech.

These games aren't just fun; they're also engaging, and they improve job performance and retention. Traci Sitzmann, a management professor at the University of Colorado, studies the role of games in on-the-job training. In one sweeping study, she examined the results from sixty-five studies that compared game-based and offline training. Across nearly seven thousand trainees, she found that game-based training was far more effective than offline training: trainees who used video games had a 9 percent higher retention rate, remembered 11 percent more facts, and rated 14 percent higher on skill-based knowledge tests. Trainees also felt 20 percent more confident and capable after playing the games, since they relied on active, hands-on experience rather than passive instruction.

The same properties that make training engaging and palatable can also be harnessed for medical benefits. In 1996, a team of researchers at the University of Washington, in Seattle, received a government grant to study the effects of virtual reality gaming on pain tolerance. Burn victims are forced to endure awful pain on a daily basis, when their wounds are cleaned and their dressing replaced. In one study, 86 percent of all burns patients described their pain levels as "excruciating," and this was *after* they were given morphine to treat the pain.

Some of the lab's patients responded well to hypnosis, so the researchers designed a virtual-reality game called SnowWorld. A distraction like SnowWorld is critical because much of a patient's pain comes from anticipation. As the researchers explain on their website:

> Our logic for why virtual reality will reduce pain is as follows. Pain perception has a strong psychological component.

The same incoming pain signal can be interpreted as painful or not, depending on what the patient is thinking. Pain requires conscious attention. The essence of VR is the illusion users have of going inside the computer-generated environment. Being drawn into another world drains a lot of attentional resources, leaving less attention available to process pain signals. Conscious attention is like a spotlight. Usually it is focused on the pain and wound care. We are luring that spotlight into the virtual world. Rather than having pain as the focus of their attention, for many patients in VR, the wound care becomes more of an annoyance, distracting them from their primary goal of exploring the virtual world.

SnowWorld is a first-person virtual reality adventure game. Players throw snowballs at penguins, mastodons, and snowmen while listening to upbeat songs by Paul Simon. The experience is immersive, and some burn patients describe playing the game as "fun"—a long way from the "excruciating" label they gave to the process of having their burns dressed before playing the game. When the researchers scanned patients' brains, they found that pain regions were less active when the patients were playing SnowWorld than when they relied on morphine alone. The same process works for other painful experiences, too—the researchers have shown that it reduces dental pain, pain suffered by children as well as adults, and the psychiatric trauma of survivors of the World Trade Center attacks on September 11, 2001.

Though Tetris, which I first discussed in chapter 7, is wickedly addictive, it also shares some of SnowWorld's therapeutic properties. Many people who witness death, injury, or a threat to others suffer ongoing trauma. The scenes they're exposed to play on a loop, over and over, sometimes for the rest of their lives. Therapists have tools to treat post-traumatic stress disorder (PTSD), but those tools aren't very effective immediately after the traumatic event. For a few weeks, the standard therapeutic approach doesn't do much good, so survivors are generally forced to wait for treatment. This struck a team of psychiatrists at the University of Oxford as strange: why wait for the memories to crystallize before beginning treatment?

In 2009, the team, led by Emily Holmes, tested a novel PTSD intervention. They asked a group of adults to watch a twelve-minute video featuring "eleven clips of traumatic content including graphic real scenes of human surgery, fatal road traffic accidents and drowning." This was their trauma simulation, and the participants who completed their study were indeed traumatized. Before the intervention they reported feeling calm and relaxed; afterward they were disturbed and jittery. Holmes and her team forced the adults to wait for thirty minutes—a simulation of the half hour wait a person might experience before being admitted to an emergency room. Then, half the participants played Tetris for ten minutes, while the other half sat quietly.

The adults went home for a week, and recorded their thoughts in a daily diary. Once a day they recounted the scenes from the video that replayed in their minds. Some saw cars colliding, and others remembered horrific scenes of human surgery. But the flashbacks affected some people more than others. Those who had sat quietly after watching the harrowing video experienced an average of six flashbacks during the week; those who had played Tetris experienced an average of fewer than three. Tetris, with its colors and music and rotating blocks, prevented the initial traumatic memories from solidifying. The game soaked up the mental attention that might have otherwise moved those horrific memories to long-term memory, and so they were stored imperfectly or not at all. At the end of the week, the adults returned to the lab, and those who had been lucky enough to play Tetris reported fewer psychiatric symptoms. The game had functioned as a "cognitive vaccine," the researchers explained. Although the video had traumatized them in the short-term, Tetris had prevented it from traumatizing them in the long-term.

Gamification is widely celebrated, but it also has detractors. In 2013, a large team of researchers published a paper on games in *Nature*, one of the world's premier science journals. The paper praised a game called NeuroRacer, which required players to steer a car while pushing buttons in response to on-screen prompts. This form of multitasking, the authors argued, was deeply therapeutic for older adults. Instead of declining, their mental functioning would remain sharp if they played NeuroRacer for an hour, three times a week. That isn't much to ask in exchange for staving off mental decline. The authors asked almost two hundred adults to play the game for a month, and then measured their mental performance for six months. Compared to older adults who didn't play at all, or who played a simpler version of the game, those who played the multitasking version did better on a wide battery of cognitive tests.

Several brain-training software companies emerged in the

wake of this research. They earned billions of dollars in revenue promoting the idea that multitasking games would improve broader mental functioning. But the evidence was scattered. Some researchers replicated the *Nature* findings, but others argued that brain-training only improved performance on trivial games; it couldn't actually improve people's lives in the long run, years and decades beyond the scope of the original experiment. In 2014, seventy-five scientists signed a statement concluding "there is no compelling scientific evidence to date" that brain games can prevent cognitive decline. The Federal Trade Commission seemed to agree. In January 2016, the FTC fined Lumos Labs, one of the largest and most successful brain-training companies, $2 million. According to the FTC, Lumos had engaged in "deceptive advertising" of its software. It was possible that Lumos's games warded off cognitive decline, but the evidence was scant, and Lumos had overclaimed.

Even if gamification works, some critics believe it should be abandoned. Ian Bogost, a game designer at Georgia Tech, spearheads this movement. In 2011, he delivered a talk at a gamification symposium at Wharton. He titled his talk "Gamification Is Bullshit." Bogost suggested that gamification "was invented by consultants as a means to capture the wild, coveted beast that is video games and to domesticate it." Bogost criticized gamification because it undermined the "gamer's" well-being. At best, it was indifferent to his well-being, pushing an agenda that he had little choice but to pursue. Such is the power of game design: a well-designed game fuels behavioral addiction.

Bogost demonstrated the power of gamification with a social media game called Cow Clicker. He designed Cow Clicker to mimic similar games, like FarmVille, which had dominated Facebook for many months. The game's objective was simple: click your cow during critical periods and you'll earn virtual currency known as *mooney*. Cow Clicker was supposed to satirize gamification, but it was a smash hit. Tens of thousands of users downloaded the game, and instead of playing once or twice, they played for days on end. At one point, a computer science professor sat atop the leaderboard with a hundred thousand mooney. Bogost updated the game with new features, adding awards for reaching certain milestones (such as the Golden Cowbell for one hundred thousand clicks), and introducing an oil-coated cow to commemorate the BP oil spill. He claimed that Cow Clicker's success was a surprise, but really it embodied many of the traits that made other games addictive: Werbach and Hunter's points, badges, and levels.

On one level, Cow Clicker is harmless fun. But Bogost makes an important point when he says that not everything should be a game. Take the case of a young child who prefers not to eat. One option is to turn eating into a game—to fly the food into his mouth like an airplane. That makes sense right now, maybe, but in the long run the child sees eating as a game. It takes on the properties of games: it must be fun and engaging and interesting, or else it isn't worth doing. Instead of developing the motivation to eat because food is sustaining and nourishing, he learns that eating is a game.

In truth, it probably doesn't matter much whether the child thinks of eating as a game or not. He'll learn the purpose of eating soon enough. But just as he replaced eating's true motive with fun, so gamification trivializes other experiences. The piano stairs at Odenplan are a lot of fun, but they don't actually promote healthy behavior in the long run. Instead, they might undermine it by suggesting that working out should be fun, primarily, rather than designed to instill health and well-being. Cute gamified interventions like the piano stairs are charming, but they're unlikely to change how people approach exercise tomorrow, next week, or next year.

In fact, the fun in gamification may crowd out important motives by changing how people see the experience entirely. In the late 1990s, economists Uri Gneezy and Aldo Rustichini tried to discourage parents from showing up late to collect their children from ten Israeli day care centers. The rational economic approach is to punish people when they're doing the wrong thing, so some of the day care centers began fining parents who showed up late. At the end of each month, their day care bills reflected these fines—an attempt to dissuade them from showing up late the following month. In fact, the fines had the opposite effect. Parents at the day care centers with fines showed up late *more* often than did parents at the day care centers without fines. The problem, Gneezy and Rustichini explained, was that the fines crowded out the motive to do the right thing. Parents felt bad coming late—until coming late became a matter of money. Then, instead of feeling bad, they saw coming late as an economic decision. The intrinsic motive to do good—to show up on time—was crowded out by the extrinsic motive to show up late in exchange for a fair price. The same is true of gamification: people think about the experience differently as soon as it adopts the hallmarks of fun. Now exercising isn't about being healthy; it's about having fun. And as soon as the fun ends, so will the exercise.

Gamification is a powerful tool, and like all powerful tools it brings mixed blessings. On the one hand, it infuses mundane or unpleasant experiences with a measure of joy. It gives medical patients respite from pain, schoolkids relief from boredom, and gamers an excuse to donate to the needy. By merely raising the number of good outcomes in the world, gamification has value. It's a worthwhile alternative to traditional medical care, education, and charitable giving because, in many respects, those approaches are tone-deaf to the drivers of human motivation. But Ian Bogost was also wise to illuminate the dangers of gamification. Games like FarmVille and Kim Kardashian's Hollywood are designed to exploit human motivation for financial gain. They pit the wielder of gamification in opposition to the gamer, who becomes ensnared in the game's irresistible net. But, as I mentioned early in this book, tech is not inherently good or bad. The same is true of gamification. Stripped of its faddish popularity and buzzwordy name, the heart of gamification is just an effective way to design experiences. Games just happen to do an excellent job of relieving pain, replacing boredom with joy, and merging fun with generosity.

Epilogue

Half of the developed world is addicted to something, and for most people that something is a behavior. We're hooked on our phones and email and video games and TV and work and shopping and exercise and a long list of other experiences that exist on the back of rapid technological growth and sophisticated product design. Few of those experiences existed in the year 2000, and by the year 2030 we'll be grappling with a new list that barely overlaps with the current roster. What we do know is that the number of immersive and addictive experiences is rising at an accelerating rate, so we need to understand how, why, and when people first develop and then escape behavioral addictions. On the lofty end of the spectrum, our health, happiness, and well-being depend on it—and right here, down to earth, so does our ability to look one another in the eyes to form genuine emotional connections.

When adults look back on the past, they tend to feel that much has changed. Things move faster than they used to; we used to talk more; times were simpler once; and so on. Despite the sense that things have changed in the past, we also tend to believe that they'll stop changing—that we and the lives we lead right now will remain this way forever. This is known as the *end of history* illusion, and it happens in part because it's much easier to see the real changes between ten years ago and today than it is to imagine how much different things will be ten years in the future. The illusion is comforting, in a way, because it makes us feel that we've finished becoming who we are, and that life will remain as it is forever. At the same time, it prevents us from preparing for the changes that are yet to come.

This is certainly true of behavioral addiction, which seems to have reached a peak. A decade ago, who could have imagined that Facebook would attract 1.5 billion users, many of whom say they wished they spent less time on the site? Or that millions of Instagram users would spend hours uploading and liking the sixty million new photos the app hosts every day? Or that more than twenty million people would count and monitor their every step with a small wrist-bound device?

These are remarkable statistics, but they represent an early waypoint on a long climb. Behavioral addiction is still in its infancy, and there's a good chance we're still at base camp, far below the peak. Truly immersive experiences, like virtual reality devices, have not yet gone mainstream. In ten years, when all of us own a pair of virtual reality goggles, what's to keep us tethered to the real world? If human relationships suffer in the face of smartphones and tablets, how are they going to withstand the tide of immersive virtual reality experiences? Facebook is barely a decade old, and Instagram is half that; in ten years, a host of new platforms will make Facebook and Instagram seem like ancient curiosities. They may still attract a large user base—it pays to launch early—but perhaps they'll be gen-one relics that have a fraction of the immersive power of the latest generation of alternatives. Of course we don't know exactly how the world will look in ten years, but, looking back on the past decade, there's no reason to believe that history has ended today, and that behavioral addiction has peaked with Facebook, Instagram, Fitbit, and World of Warcraft.

So what's the solution? We can't abandon technology, nor should we. Some technological advances fuel behavioral addiction, but they are also miraculous and life enriching. And with careful engineering they don't need to be addictive. It's possible to create a product or experience that is indispensable but not addictive. Workplaces, for example, can shut down at six—and with them work email accounts can be disabled between midnight and five the next morning. Games, like books with chapters, can be built with natural stopping points. Social media platforms can "demetricate," removing the numerical feedback that makes them vehicles for damaging social comparison and chronic goal-setting. Children can be introduced to screens slowly and with supervision, rather than all at once. Our attitude to addictive experiences is largely cultural, and if our culture makes space for work-free, game-free, screen-free downtime, we and our children will find it easier to resist the lure of behavioral addiction. In its place, we'll communicate with one another directly, rather than through devices, and the glow of these social bonds will leave us richer and happier than the glow of screens ever could.

Acknowledgments

A huge thank you to the teams at Penguin Press, Inkwell Management, and Broadside PR. At Penguin Press, in particular, to my wise and patient editor, Ann Godoff, who made *Irresistible* far stronger and tighter than I could have managed alone. Also at Penguin Press, thanks to Will Heyward, Juliana Kiyan, Sara Hutson, Matt Boyd, Caitlin O'Shaughnessy, and Casey Rasch. At Inkwell, special thanks to my kind, insightful agent, Richard Pine, who is everything an agent should be: an ideas man, a psychologist, a publicity guru, and a friend. Also at Inkwell, thanks to Eliza Rothstein and Alexis Hurley. And at Broadside, thanks to Whitney Peeling and the entire Broadside team.

For reading earlier drafts of *Irresistible*, sharing their ideas, and patiently answering my questions, thanks to Nicole Airey, Dean Alter, Jenny Alter, Ian Alter, Sara Alter, Chloe Angyal, Gary Aston Jones, Nicole Avena, Jessica Barson, Kent Berridge, Michael Brough, Oliver Burkeman, Hilarie Cash, Ben Caunt, Rameet Chawla, John Disterhoft, Andy Doan, Natasha Dow Schüll, David Epstein, Bennett Foddy, Allen Frances, Claire Gillan, Malcolm Gladwell, David Goldhill, Adam Grant, Melanie

Green, Mark Griffiths, Hal Hershfield, Jason Hirschel, Kevin Holesh, Margot Lacey, Frank Lantz, Andrew Lawrence, Tom Meyvis, Stanton Peele, Jeff Peretz, Ryan Petrie, Sam Polk, Cosette Rae, Aryeh Routtenberg, Adam Saltsman, Katherine Schreiber, Maneesh Sethi, Eesha Sharma, Leslie Sim, Anni Sternisko, Abby Sussman, Maia Szalavitz, Isaac Vaisberg, Carrie Wilkens, Bob Wurtz, and Kimberly Young.

In late 2014, I described the premise of *Irresistible* to the students who took my marketing class at New York University's Stern School of Business. Thanks to those who helped me by sending anecdotes and examples of addictive tech, particularly Griffin Carlborg, Caterina Cestarelli, Gizem Ceylan, Arianna Chang, Jane Chyun, Sanhita Dutta Gupta, Elina Hur, Allega Ingerson, Nishant Jain, Chakshu Madhok, Danielle Nir, Michelle See, Yash Seksaria, Yu Sheng, Jenna Steckel, Sonya Shah, Lindsay Stecklein, Anne-Sophie Svoboda, Madhumitha Venkataraman, and Amy Zhu.

And thanks, always, to my wife, Sara; my son, Sam; my parents, Ian and Jenny; Suzy and Mike; and my brother Dean.

Notes

PROLOGUE: NEVER GET HIGH OR YOUR OWN SUPPLY

1   **At an Apple event:** John D. Sutter and Doug Gross, "Apple Unveils the 'Magical' iPad," CNN, January 28, 2010, www.cnn.com/2010/TECH/01/27/apple.tablet/. Video of the event: EverySteveJobsVideo, "Steve Jobs Introduces Original iPad—Apple Special Event," December 30, 2013, www.youtube.com/watch?v=_KN-5zmvjAo.

2   **In late 2010, Jobs:** This section of views from tech experts comes from: Nick Bilton, "Steve Jobs Was a Low-Tech Parent," *New York Times*, September 11, 2014, www.nytimes.com/2014/09/11/fashion/steve-jobs-apple-was-a-low-tech-parent.html.

2   **Many experts both:** These snippets come from interviews with, among others, game designers Bennett Foddy and Frank Lantz, exercise addiction experts Leslie Sim and Katherine Schreiber, and reSTART Internet addiction clinic founder Cosette Rae.

3   **Greg Hochmuth, one:** These quotes are from: Natasha Singer, "Can't Put Down Your Device? That's by Design," *New York Times*, December 5, 2015, www.nytimes.com/2015/12/06/technology/personaltech/cant-put-down-your-device-thats-by-design.html.

5   **Tech offers convenience:** For more on how technology-aided speed drives behavioral addiction, see: Art Markman, "How to Disrupt Your Brain's Distraction Habit," Inc.com, May 25, 2016, www.inc.com/art-markman/the-real-reason-technology-destroys-your-attention-span-is-timing.html.

5   **These new addictions:** For the purposes of this book, I adopted my own definitions of behavioral addiction, compulsion, and obsession, which borrowed from several sources. In particular, I relied on the following handbook, an accessible scholarly work on behavioral addiction that gathers chapters from dozens of experts: Kenneth Paul Rosenberg and Laura Curtiss Feder, eds., *Behavioral Addictions: Criteria, Evidence, and Treatment* (Elsevier Academic Press: London, 2014). I also relied on: Aviel Goodman, "Addiction: Definitions and Implications," *British Journal of Addiction* no. 85 (1990): 1403-8. To some extent, I adopted the definitions in: American Psychiatric Association, *Diagnostic and Statistical Manual of Mental Disorders*, 5th ed. (American Psychiatric Publishing: Washington, DC, 2013).

6   **I spoke to several:** These clinical psychologists agreed to speak on condition that I refrained from using their names. They were concerned their patients might recognize the anecdotes they relayed anonymously.

8   **Writing for *Time*:** John Patrick Pullen, "I Finally Tried Virtual Reality and It Brought Me to Tears," *Time*, January 8, 2016, www.time.com/4172998/virtual-reality-oculus-rift-htc-vive-ces/.

CHAPTER 1: THE RISE OF BEHAVIORAL ADDICTION

13 **On the Moment:** The Moment website: inthemoment.io/; Holesh's blog: inthemoment.io/blog. Other pieces on Holesh and his app include: Conor Dougherty, "Addicted to Your Phone? There's Help for That," *New York Times*, July 11, 2015, www.nytimes.com/2015/07/12/sunday-review/addicted-to-your-phone-theres-help-for-that.html; Seth Fiegerman, "'You've Been on Your Phone for 160 Minutes Today,'" Mashable, August 14, 2014, mashable.com/2014/08/19/mobile-addiction/; Sarah Perez, "A New App Called Moment Shows You How Addicted You Are to Your iPhone," TechCrunch, June 27, 2014, techcrunch.com/2014/06/27/a-new-app-called-moment-shows-you-how-addicted-you-are-to-your-iphone/; Jiaxi Lu, "This App Tells You How Much Time You Are Spending, or Wasting, on Your Smartphone," *Washington Post*, August 21, 2014, www.washingtonpost.com/news/technology/wp/2014/08/21/this-app-tells-you-how-much-time-you-are-spending-or-wasting-on-your-smartphone/.

15 **This sort of overuse:** Research on the topic includes: ALS. King and others, "Nomophobia: Dependency on Virtual Environments or Social Phobia?," *Computers in Human Behaviors* 29, no. 1 (January 2013): 140–44; A. L. S. King, A. M. Valença, and A. E. Nardi, "Nomophobia: The Mobile Phone in Panic Disorder with Agoraphobia: Reducing Phobias or Worsening of Dependence?," *Cognitive and Behavioral Neurology* 23, no. 1 (2010): 52–54; James A. Roberts, Luc Honore Petnji Yaya, and Chris Manolis, "The Invisible Addiction: Cell-Phone Activities and Addiction Among Male and Female College Students," *Journal of Behavioral Addictions* 3, no. 4 (December 2014): 254–65; Andrew Lepp, Jacob E. Barkley, and Aryn C. Karpinski, "The Relationship between Cell Phone Use, Academic Performance, Anxiety, and Satisfaction with Life in College Students," *Computers in Human Behavior* 31 (February 2014) 343–50; Shari P. Walsh, Katherine M. White, Ross McD. Young, "Needing to Connect: The Effect of Self and Others on Young People's Involvement with Their Mobile Phones," *Australian Journal of Psychology* 62, no. 4 (2010): 194–203.

15 **damaging. In 2013:** Andrew K. Przybylski and Netta Weinstein, "Can You Connect with Me Now? How the Presence of Mobile Communication Technology Influences Face-to-Face Conversation Quality," *Journal of Social and Personal Relationships* 30, no. 3 (May 2013): 237–46.

16 **WoW may be:** Colin Lecher, "GameSci: What Is (Scientifically!) the Most Addictive Game Ever?," *Popular Science*, March 27, 2013, www.popsci.com/gadgets/article/2013-03/gamesci-what-scientifically-most-addictive-game-ever; WoWaholics Anonymous discussion board, www.reddit.com/r/nowow/; WoW Addiction Test, www.helloquizzy.com/tests/the-new-and-improved-world-of-warcraft-addiction-test.

17 **the game has:** Ana Douglas, "Here Are the 10 Highest Grossing Video Games Ever," Business Insider, June 13, 2012, www.businessinsider.com/here-are-the-top-10-highest-grossing-video-games-of-all-time-2012-6; Samit Sarkar, "Blizzard Reaches 100M Lifetime World of Warcraft Accounts," Polygon, January 28, 2014, www.polygon.com/2014/1/28/5354856/world-of-warcraft-100m-accounts-lifetime.

17 **up to 40 percent—develop:** Jeremy Reimer, "Doctor Claims 40 Percent of World of Warcraft Players Are Addicted," *Ars Technica*, August 9, 2006, arstechnica.com/uncategorized/2006/08/7459/.

17 **The center, named:** Information on reSTART: www.netaddictionrecovery.com/.

19 **The first "behavioral:** Jerome Kagan, "The Distribution of Attention in Infancy," in *Perception and Its Disorders*, eds. D. A. Hamburg, K. H. Pribram, and A. J. Stunkard, (Williams and Wilkins Company: Baltimore, MD, 1970), 214–37.

21 **Behavioral addiction also:** R. J. Vallerand and others, "Les passions de l'ame: On Obsessive and Harmonious Passion," *Journal of Personality and Social Psychology* 83 (2003): 756–67.

23   **Still, it's important:** For more on Allen Frances's views, see: Allen Frances, "Do We All Have Behavioral Addictions?," *Huffington Post*, March 28, 2012, www.huffingtonpost .com/allen-frances/behavioral-addiction_b_1215967.html.

24   **Just how common:** Steve Sussman, Nadra Lisha, and Mark D. Griffiths, "Prevalence of the Addictions: A Problem of the Majority or the Minority?," *Evaluation and the Health Professions* 34 (2011): 3–56.

26   **One recent study suggested that up:** These statistics come from: Susan M. Snyder, Wen Li, Jennifer E. O'Brien, and Matthew O. Howard, "The Effect of U.S. University Students' Problematic Internet Use on Family Relationships: A Mixed-methods Investigation," *Plos One*, December 11, 2015, journals.plos.org/plosone/article?id=10.1371/journal.pone.0144005.

26   **This is a sample:** You can find the complete IAT here: netaddiction.com/Internet-addiction-test/.

27   **46 percent of:** All statistics here contained in Rosenberg and Feder, *Behavioral Addictions*. See also: Aaron Smith, "U.S. Smartphone Use in 2015," PewResearchCenter, April 1, 2015, www. pewInternet.org/2015/04/01/us-smartphone-use-in-2015/; Ericsson Consumer Lab, "TV and Media 2015: The Empowered TV and Media Consumer's Influence," September 2015.

28   **and 80 percent:** Kelly Wallace, "Half of Teens Think They're Addicted to their Smart-phones," CNN, May 3, 2016, www.cnn.com/2016/05/03/health/teens-cell-phone-addiction -parents/index.html.

28   **In 2008, adults:** Kleiner Perkins Caulfield & Byers, "Internet Trends Report 2016," SlideShare, May 26, 2015, www.slideshare.net/kleinerperkins/internet-trends-v1/14-14In ternet_Usage_Engagement_Growth_Solid11.

28   **In 2000, Microsoft:** Microsoft Canada, Consumer Insights, *Attention Spans*, Spring 2015, advertising.microsoft.com/en/WWDocs/User/display/cl/researchreport/31966/en/ microsoft-attention-spans-research-report.pdf. Microsoft couldn't conclude with certainty that social media compromised attention. It was possible, for example, that the kinds of people who use social media are less attentive in general. But in concert with the report's other findings, the correlation was concerning.

29   **Addiction originally meant:** Etymology of "addiction": *Oxford English Dictionary*, 1989, www.oup.com; see also Mark Peters, "The Word We're Addicted To," CNN, March 23, 2010, www.cnn.com/2011/LIVING/03/23/addicted.to.addiction/.

30   **DNA evidence suggests:** Justin R. Garcia and others, "Associations Between Dopamine D4 Receptor Gene Variation with Both Infidelity and Sexual Promiscuity," *Plos One*, 2010, journals.plos.org/plosone/article?id=10.1371/journal.pone.0014162; see also: B. P. Zietsch and others, "Genetics and Environmental Influences on Risky Sexual Behaviour and Its Relationship with Personality," *Behavioral Genetics* 40, no. 1 (2010): 12–21; David Cesarini and others, "Genetic Variation in Financial Decision-making," *The Journal of Finance* 65, no. 5 (October 2010): 1725–54; David Cesarini and others, "Genetic Variation in Preferences for Giving and Risk Taking," *Quarterly Journal of Economics* 124, no. 2 (2009): 809–42; Songfa Zhong and others, "The Heritability of Attitude Toward Economic Risk," *Twin Research and Human Genetics* 12, no. 1 (2009): 103–7.

30   **he or she lived:** See, for example, Tammy Saah, "The Evolutionary Origins and Significance of Drug Addiction," *Harm Reduction Journal* 2, no 8 (2005), harmreduction journal.biomedcentral.com/articles/10.1186/1477-7517-2-8.

30   **The betel nut:** History of addictions from: Jonathan Wynne-Jones, "Stone Age Man Took Drugs, Say Scientists," *Telegraph*, October 19, 2008, www.telegraph.co.uk/news/newstopics/ howaboutthat/3225729/Stone-Age-man-took-drugs-say-scientists.html; Marc-Antoine Crocq, "Historical and Cultural Aspects of Man's Relationship with Addictive Drugs," *Dialogues in Clinical Neuroscience* 9, no. 4 (2007): 355–61; Tammy Saah, "The Evolutionary Origins

and Significance of Drug Addiction," *Harm Reduction Journal* 2, no. 8 (2005) harm reductionjournal.biomedcentral.com/articles/10.1186/1477-7517-2-8; Nguyên Xuân Hiên, "Betel-Chewing in Vietnam: Its Past and Current Importance," *Anthropos* 101 (2006): 499–516; Hilary Whiteman, "Nothing to Smile About: Asia's Deadly Addiction to Betel Nuts," CNN, November 5, 2013, www.cnn.com/2013/11/04/world/asia/myanmar-betel-nut-cancer.

32 **In 1875 the:** David F. Musto, "America's First Cocaine Epidemic," *The Wilson Quarterly* 13, no. 3 (Summer 1989): 59–64; Curtis Marez, *Drug Wars: The Political Economy of Narcotics* (Minneapolis: University of Minnesota Press, 2004); Robert Christison, "Observations on the Effects of the Leaves of Erythroxylon Coca," *British Medical Journal* 1 (April 29, 1876): 527–31.

33 **It began with:** A good summary of Freud and "Über Coca": "Über Coca, by Sigmund Freud," scicurious, May 28, 2008, scicurious.wordpress.com/2008/05/28/uber-coca-by-sigmund-freud/; Sigmund Freud, "Über Coca" classics revisited, *Journal of Substance Abuse and Treatment* 1 (1984), 206–17; Howard Markel, *An Anatomy of Addiction: Sigmund Freud, William Halsted, and the Miracle Drug, Cocaine* (New York: Vintage 2012).

37 **Like any good:** On Pemberton and Coca-Cola: Bruce S. Schoenberg, "Coke's the One: The Centennial of the 'Ideal Brain Tonic' That Became a Symbol of America," *Southern Medical Journal* 81, no. 1 (1988): 69–74; M. M. King, "Dr. John S. Pemberton: Originator of Coca-Cola," *Pharmacy in History* 29, no. 2 (1987): 85–89; Guy R. Hasegawa, "Pharmacy in the American Civil War," *American Journal of Health-System Pharmacy* 57, no. 5 (2000): 457–89; Richard Gardiner, "The Civil War Origin of Coca-Cola in Columbus, Georgia," *Muscogiana: Journal of the Muscogee Genealogical Society* 23 (2012): 21–24; Dominic Streatfeild, *Cocaine: An Unauthorized Biography* (London: Macmillan, 2003); Richard Davenport-Hines, *The Pursuit of Oblivion: A Global History of Narcotics* (New York: Norton, 2004).

39 **In 2013, a psychologist:** Catherine Steiner-Adair, *The Big Disconnect: Protecting Childhood and Family Relationships in the Digital Age* (New York: Harper, 2013).

39 **using head-mounted:** Chen Yu and Linda B. Smith. "The Social Origins of Sustained Attention in One-year-old Human Infants," *Current Biology* 26, no. 9 (May 9, 2016): 1235–40.

40 **According to the paper's:** Indiana University, "Infant Attention Span Suffers When Parents' Eyes Wander During Playtime: Eye-tracking Study First to Suggest Connection between Caregiver Focus and Key Cognitive Development Indicator in Infants," ScienceDaily, April 28, 2016, www.sciencedaily.com/releases/2016/04/160428131954.htm.

41 **Like Steiner-Adair:** Nancy Jo Sales, *American Girls: Social Media and the Secret Lives of Teenagers* (New York: Knopf, 2016).

42 **Echoing Sales' account:** Jessica Contrera. "13, Right Now," *Washington Post*, May 25, 2016, www.washingtonpost.com/sf/style/wp/2016/05/25/2016/05/25/13-right-now-this-is-what-its-like-to-grow-up-in-the-age-of-likes-lols-and-longing/.

42 **In May 2013:** On Dong Nguyen and Flappy Bird: Much of the information in this section is from the original Flappy Bird download page, which is no longer available online. Other references include: John Boudreau and Aaron Clark, "Flappy Bird Creator Dong Nguyen Offers Swing Copters Game," Bloomberg Technology, August 22, 2014, www.bloomberg.com/news/articles/2014-08-22/flappy-bird-creator-dong-nguyen-offers-swing-copters-game; Laura Stampler, "Flappy Bird Creator Says 'It's Gone Forever'," *Time*, February 11, 2014, http://time.com/6217/flappy-bird-app-dong-nguyen-addictive/; James Hookway, "Flappy Bird Creator Pulled Game Because It Was 'Too Addictive,'" *Wall Street Journal*, February 11, 2014, www.wsj.com/articles/SB10001424052702303874504579376323271110900; Lananh Nguyen, "Flappy Bird Creator Dong Nguyen Says App 'Gone Forever' Because It Was "An Addictive Product,'" *Forbes*, February 11, 2014, www.forbes.com/sites/lananhnguyen/2014/02/11/exclusive-flappy-bird-creator-dong-nguyen-says-app-gone-forever-because-it-was-an-addictive-product/.

44 **Just recently a:** Kathryn Yung and others, "Internet Addiction Disorder and Problematic

Use of Google Glass in Patient Treated at a Residential Substance Abuse Treatment Program," *Addictive Behaviors* 41 (2015): 58–60; James Eng, "Google Glass Addiction? Doctors Report First Case of Disorder," NBC News, October 14, 2014, www.nbcnews.com/tech/Internet/google-glass-addiction-doctors-report-first-case-disorder-n225801.

CHAPTER 2: THE ADDICT IN ALL OF US

46 **Most war films:** Jason Massad, "Vietnam Veteran Recalls Firefights, Boredom and Beer," *Reporter Newspapers*, November 4, 2010, www.reporternewspapers.net/2010/11/04/vietnam -veteran-recalls-firefights-boredom-beer/.

46 **Vietnam lies just:** Background on the Golden Triangle heroin trade during the Vietnam War, and Nixon's response: Alfred W. McCoy, Cathleen B. Read, and Leonard P. Adams II, *The Politics of Heroin in Southeast Asia* (New York: Harper and Row, 1972); Tim O'Brien, *The Things They Carried* (New York: Houghton Mifflin Harcourt, 1990); Liz Ronk, "The War Within: Portraits of Vietnam War Veterans Fighting Heroin Addiction, *Time*, January 20, 2014, time.com/3878718/vietnam-veterans-heroin-addiction-treatment-photos/; Aimee Groth, "This Vietnam Study about Heroin Reveals the Most Important Thing about Kicking Addictions," Business Insider, January 3, 2012, www.businessinsider. com/vietnam-study-addictions-2012-1; Dirk Hanson, "Heroin in Vietnam: The Robins Study," Addiction Inbox, July 24, 2010, addiction-dirkh.blogspot.com/2010/07/heroin-in -viet-nam-robins-study.html; Jeremy Kuzmarov, *The Myth of the Addicted Army: Vietnam and the Modern War on Drugs* (Amherst, MA: University of Massachusetts Press, 2009); Alix Spiegel, "What Vietnam Taught Us about Breaking Bad Habits," NPR, January 2, 2012, www.npr.org/sections/health-shots/2012/01/02/144431794/what-vietnam-taught-us-about-breaking-bad-habits; Alexander Cockburn and Jeffrey St. Clair, *Whiteout: The CIA, Drugs, and the Press*, (New York: Verso, 1997).

48 **When British researchers:** David Nutt, Leslie A. King, William Saulsbury, and Colin Blakemore, "Development of a Rational Scale to Assess the Harm of Drugs of Potential Misuse," *Lancet* 369, no. 9566 (March 2007): 1047–53.

49 **In Vietnam, Major:** Peter Brush, "Higher and Higher: American Drug Use in Vietnam," *Vietnam Magazine*, December 2002, nintharticle.com/vietnam-drug-usage.htm; Alfred W. McCoy, Cathleen B. Read, and Leonard P. Adams II, *The Politics of Heroin in Southeast Asia* (New York: Harper and Row, 1972).

50 **At home, the:** Background on Lee Robins, and her own reports: Lee N. Robins, "Vietnam Veterans' Rapid Recovery from Heroin Addiction: A Fluke or Normal Expectation?," *Addiction* 88, no. 8 (1993), 1041–54; Lee N. Robins, John E. Helzer, and Darlene H. Davis, "Narcotic Use in Southeast Asia and Afterward," *Archives of General Psychiatry* 32, no. 8 (1975): 955–961; Lee N. Robins and S. Slobodyan, "Post-Vietnam Heroin Use and Injection by Returning US Veterans: Clues to Preventing Injection Today," *Addiction* 98, no. 8 (2003): 1053–60; Lee N. Robins, Darlene H. Davis, and Donald W. Goodwin, "Drug Use by U.S. Army Enlisted Men in Vietnam: A Follow-up on Their Return Home," *American Journal of Epidemiology* 99, no. 4 (May 1974): 235–49; Lee N. Robins, *The Vietnam Drug User Returns*, final report, Special Action Office Monograph, Series A, Number 2, May 1974, prhome.defense.gov/Portals/52/Documents/RFM/Readiness/DDRP/docs/35%20 Final%20Report.%20The%20Vietnam%20drug%20user%20returns.pdf; Lee N. Robins, John E. Helzer, Michie Hesselbrock, and Eric Wish, "Vietnam Veterans Three Years after Vietnam: How Our Study Changed Our View of Heroin," *American Journal on Addictions* 19, 203–11 (2010); Thomas H. Maugh II, "Lee N. Robins Dies at 87; Pioneer in Field of Psychiatric Epidemiology," *Los Angeles Times*, October 6, 2009, www.latimes.com/nation/la-me-lee-robins6-2009oct06-story.html.

52 **If the engineer:** Information on Olds and Milner comes from two sources—interviews with their students: Bob Wurtz, Gary Aston-Jones, Aryeh Routtenberg, and John Disterhoft;

and various written resources: James Olds and Peter Milner, "Positive Reinforcement Produced by Electrical Stimulation of Septal Area and Other Regions of Rat Brain," *Journal of Comparative and Physiological Psychology* 47, no. 6 (December 1954): 419–27; James Olds, "Pleasure Centers in the Brain," *Scientific American* 195 (1956): 105–16; James Olds and M. E. Olds, "Positive Reinforcement Produced by Stimulating Hypothalamus with Iproniazid and Other Compounds," *Science* 127, no. 3307 (May 16, 1958): 1155–56; Robert H. Wurtz, *Autobiography*, n.d., www.sfn.org/~/media/SfN/Documents/TheHistoryofNeuroscience/Volume%207/c16.ashx; Richard F. Thompson, *James Olds: Biography* (National Academies Press, 1999) www.nap.edu/read/9681/chapter/16.

60   **Isaac Vaisberg, a:** Background on Vaisberg, including his addiction to WoW and his affiliation with reSTART, from two interviews with Vaisberg.

CHAPTER 3: THE BIOLOGY OF BEHAVIORAL ADDICTION

68   **There's a modern-day:** Anne-Marie Chang, Daniel Aeschbach, Jeanne F. Duffy, and Charles A. Czeisler, "Evening Use of Light-emitting eReaders Negatively Affects Sleep, Circadian Timing, and Next-morning Alertness," *Proceedings of the National Academy of Sciences* 112, no. 4 (2015): 1232–37; Brittany Wood, Mark S. Rea, Barbara Plitnick, and Mariana G. Figueiro, "Light Level and Duration of Exposure Determine the Impact of Self-luminous Tablets on Melatonin Suppression," *Applied Ergonomics* 44, no. 2 (March 2013) 237–40. Apple recently introduced a function called Night Shift into its screen-based devices, which changes the color of the screen through the day to reduce blue light before bedtime: www.apple.com/ios/preview/. More on this: Margaret Rhodes, "Amazon and Apple Want to Save Your Sleep by Tweaking Screen Colors," *Wired*, January 1, 2016, www.wired.com/2016/01/amazon-and-apple-want-to-improve-your-sleep-by-tweaking-screen-colors/; TechCrunch, "Arianna Huffington on Technology Addiction and the Sleep Revolution," January 20, 2016, techcrunch.com/video/arianna-huffington-on-politics-and-her-new-book-the-sleep-revolution/519432319/.

70   **The human brain exhibits:** K. M. O'Craven and N. Kanwisher, "Mental Imagery of Faces and Places Activates Corresponding Stimulus-Specific Brain Regions," *Journal of Cognitive Neuroscience* 12, no. 6 (2000): 1013–23; Nancy Kanwisher, Josh McDermott, and Marvin M. Chun, "The Fusiform Face Area: A Module in Human Extrastriate Cortex Specialized for Face Perception," *Journal of Neuroscience* 17, no. 11 (June 1, 1997): 4302–311.

71   **There's also a:** Much of the information in this chapter comes from interviews with addiction and physiological psychology researchers and experts: Claire Gillan, Nicole Avena, Jessica Barson, Kent Berridge, Andrew Lawrence, Stanton Peele, and Maia Szalavitz.

74   **In one article:** Maia Szalavitz, "Most of Us Still Don't Get It: Addiction Is a Learning Disorder," *Pacific Standard*, August 4, 2014, www.psmag.com/health-and-behavior/us-still-dont-get-addiction-learning-disorder-87431; see also Maia Szalavitz, "How the War on Drugs Is Hurting Chronic Pain Patients," Vice, July 16, 2015, www.vice.com/read/how-the-war-on-drugs-is-hurting-chronic-pain-patients-716; Maia Szalavitz, "Curbing Pain Prescriptions Won't Reduce Overdoses. More Drug Treatment Will," *Guardian*. March 26, 2016, www.theguardian.com/commentisfree/2016/mar/29/prescription-drug-abuse-addiction-treatment-painkiller.

75   **In 2005, an:** Arthur Aron and others, "Reward, Motivation, and Emotion Systems Associated with Early-Stage Intense Romantic Love," *Journal of Neurophysiology* 94, no. 1 (July 1, 2005), 327–37; see also: Helen Fisher, "Love Is Like Cocaine," *Nautilus*, February 4, 2016, nautil.us/issue/33/attraction/love-is-like-cocaine. See also: Richard A. Friedman, "I Heart Unpredictable Love," *New York Times,* November 2, 2012, www.nytimes.com/2012/11/04/opinion/sunday/i-heart-unpredictable-love.html; Helen Fisher, Arthur Aron, and Lucy L. Brown, "Romantic Love: An fMRI Study of a Neural Mechanism for Mate Choice," *Journal of Comparative Neurology* 493 (2005): 58-62.

76   **In the 1970s, a psychologist:** Information on Peele comes from an interview with Peele, and three books: Stanton Peele and Archie Brodsky, *Love and Addiction* (New York: Taplinger, 1975); Stanton Peele, *The Meaning of Addiction: An Unconventional View* (Lexington, MA: Lexington Books, 1985); Stanton Peele and Archie Brodsky, with Mary Arnold, *The Truth about Addiction and Recovery: The Life Process Program for Outgrowing Destructive Habits* (New York: Fireside, 1991).

76   **In a 1990:** Isaac Marks, "Behavioural (Non-chemical) Addictions," *British Journal of Psychiatry* 85, no. 11 (November 1990): 1389–94.

80   **Every fifteen years:** American Psychiatric Association, *Diagnostic and Statistical Manual of Mental Disorders* (5th ed.), Washington, DC: American Psychiatric Publishing, 2013).

80   **In the 1960s, even:** Information on Rylander and punding from interviews with Andrew Lawrence and Kent Berridge; also see Andrew D. Lawrence, Andrew H. Evans, and Andrew J. Lees, "Compulsive Use of Dopamine Replacement in Parkinson's Disease: Reward Systems Gone Awry?," *Lancet: Neurology* 2, no. 10 (October 2003): 595–604; A. H. Evans and others, "Punding in Parkinson's Disease: Its Relation to the Dopamine Dysregulation Syndrome," *Movement Disorders* 19, no. 4 (April 2004): 397–405; Gösta Rylander, "Psychoses and the Punding and Choreiform Syndromes in Addiction to Central Stimulant Drugs," *Psychiatria, Neurologia, and Neurochirurgia* 75, no. 3 (May–June 1972): 203–12; H. H. Fernandez and J. H. Friedman, "Punding on L-Dopa," *Movement Disorders* 14, no. 5 (September 1999): 836–38; Kent C. Berridge, Isabel L. Venier, and Terry E. Robinson, "Taste reactivity analysis of 6-Hydroxydopamine-Induced Aphasia: Implications for Arousal and Anhedonia Hypotheses of Dopamine Function," *Behavioral Neuroscience* 103, no. 1 (February 1989): 36–45. Both Berridge and Lawrence have published dozens of papers on the brain and addiction; for more, see: Berridge: lsa.umich.edu/psych/research&labs/berridge/Publications.htm; Lawrence: psych.cf.ac.uk/contactsandpeople/academics/lawrence.php#publications.

83   **When Billy Connolly:** Video of Connolly discussing Parkinson's disease and his treatment on Conan O'Brien: teamcoco.com/video/billy-connolly-hobbit-hater.

87   **One recent study suggests:** Xianchi Dai, Ping Dong, Jayson S. Jia, "When Does Playing Hard to Get Increase Romantic Attraction?," *Journal of Experimental Psychology: General* 143, no. 2 (April 2014): 521–26.

CHAPTER 4: GOALS

93   **In 1987, three:** J. W. Dunne, G. J. Hankey, and R. H. Edis, "Parkinsonism: Upturned Walking Stick as an Aid to Locomotion," *Archives of Physical Medicine and Rehabilitation* 68, no. 6 (June 1987): 380–81.

96   **But that's not how the:** Eric J. Allen, Patricia M. Dechow, Devin G. Pope, and George Wu, "Reference-Dependent Preferences: Evidence from Marathon Runners," *NBER Working Paper No. 20343*, July 2014, www.nber.org/papers/w20343.

98   **Robert Beamon was:** Rob Bagchi, "50 Stunning Olympic Moments, No. 2: Bob Beamon's Great Leap Forward," *Guardian*, November 23, 2011, www.theguardian.com/sport/blog/2011/nov/23/50-stunning-olympic-bob-beamon.

100 **Larson was known:** Larson's episode on *Press Your Luck* is discussed and broadcast during a documentary titled *Big Bucks: The Press Your Luck Scandal* (James P. Taylor Jr. [director], Game Show Network, 2003); Larson's story is also recounted in: Alan Bellows, "Who Wants to Be a Thousandaire?," *Damn Interesting*, September 12, 2011, www.damninteresting.com/who-wants-to-be-a-thousandaire/; *This American Life*, "Million Dollar Idea," NPR, July 16, 2010, www.thisamericanlife.org/radio-archives/episode/412/million-dollar-idea.

107 **There's plenty of:** These searches were conducted on Google's Ngram Viewer: books.google.com/ngrams.

109 **How long do:** Thomas Jackson, Ray Dawson, and Darren Wilson, "Reducing the Effect of

Email Interruptions on Employees, *International Journal of Information Management* 23, no. 1 (February 2003): 55–65.

110 **The researchers monitored:** Information on the role of emailing at work from: Gloria J. Mark, Stephen Voida, and Armand V. Cardello, "'A Pace Not Dictated by Electrons: An Empirical Study of Work Without Email," *Proceedings of the SIGCHI Conference on Human Factors in Computer Systems* (2012): 555–64; Megan Garber, "The Latest 'Ordinary Thing That Will Probably Kill You'? Email," *The Atlantic*, May 4, 2012, www .theatlantic.com/technology/archive/2012/05/the-latest-ordinary-thing-that-will-probably -kill-you-email/256742/; Joe Pinsker, "Inbox Zero vs. Inbox 5,000: A Unified Theory," *The Atlantic*, May 27, 2015, www.theatlantic.com/technology/archive/2015/05/why-some -people-cant-stand-having-unread-emails/394031/; Stephen R. Barley, Debra E. Myerson, and Stine Grodal, "E-mail as a Source and Symbol of Stress," *Organization Science* 22, no. 4 (July–August 2011): 887–906; Mary Czerwinski, Eric Horvitz, and Susan Wilhite, "A Diary Study of Task Switching and Interruptions," *Proceedings of the Special Interest Group on Computer–Human Interaction Conference on Human Factors in Computer Systems* (2004): 175–82; Laura A. Dabbish and Robert E. Kraut, "Email Overload at Work: An Analysis of Factors Associated with Email Strain," *Proceedings of the Association for Computing Machinery Conference on Computer Supported Cooperative Work & Social Computing* (2011): 431–40; Chuck Klosterman, "My Zombie, Myself: Why Modern Life Feels Rather Undead," *New York Times*, December 3, 2010, www.nytimes.com/2010/12/05/arts/ television/05zombies.html; Karen Renaud, Judith Ramsay, and Mario Hair, "'You've Got E-Mail!' . . . Shall *I* Deal with It Now? Electronic Mail from the Recipient's Perspective," *International Journal of Human–Computer Interaction* 21, no. 3 (2006): 313–32.

112 **Katherine Schreiber and Leslie Sim are:** Information from interviews with Schreiber and Sim, and from Schreiber's book: Katherine Schreiber, *The Truth about Exercise Addiction* (New York: Rowman & Littlefield Publishers, 2015).

115 **In 2000, Marylanders:** Running Streak Association website: www.runeveryday.com/; active list of runners: www.runeveryday.com/lists/USRSA-Active-List.html; see also: Katherine Dempsey, "The People Who Can't Not Run," *The Atlantic*, June 4, 2014, www.theatlantic. com/health/archive/2014/06/streakers-in-sneakers/371347/; Kevin Helliker, "These Streak-ers Resolve to Run Every Day of the Year," *Wall Street Journal*, January 1, 2015, www.wsj. com/articles/these-streakers-resolve-to-run-every-day-of-the-year-1419986806.

117 **Writing for the:** Oliver Burkeman, "Want to Succeed? You Need Systems, Not Goals," *Guardian*, November 7, 2014, www.theguardian.com/lifeandstyle/2014/nov/07/systems -better-than—goals-oliver-burkeman. See also: Scott Adams, *How to Fail at Everything and Still Win Big: Kind of the Story of My Life* (New York: Portfolio, 2014).

118 **Sam Polk published:** Background on Polk from an interview with Polk, and from his op-ed: Sam Polk, "For the Love of Money," *New York Times*, January 14, 2014, www .nytimes.com/2014/01/19/opinion/sunday/for-the-love-of-money.html.

CHAPTER 5: FEEDBACK

121 **In 2012, an:** Turner Benelux, "A Dramatic Surprise on a Quiet Square," YouTube, April 11, 2012, www.youtube.com/watch?v=316AzLYfAzw; see also: Laura Stampler, "How TNT Made the Biggest Viral Ad of the Year—in Belgium," Business Insider, May 15, 2012, www.businessinsider.com/how-a-belgian-agency-made-one-of-the-most-viral-videos-of -this-year-2012-5; Anthony Wing Kosner), "'Push to Add Drama' Video: Belgian TNT Advert Shows Virality of Manipulated Gestures," *Forbes*, April 12, 2012, www.forbes.com/ sites/anthonykosner/2012/04/12/push-to-add-drama-video-belgian-tnt-advert-shows- virality-of-manipulated-gestures/#85072544803b.

122 **This was the case:** The archived "The Button" subreddit was still online as of May 2016: https://www.reddit.com/r/thebutton; more from Reddit's blog: www.redditblog.com

3

/2015/06/the-button-has-ended.html; see also, for example, Julianne Pepitone, "Reddit Explains the Mystery Behind 'The Button,'" NBC, June 9, 2015, www.nbcnews.com/tech/Internet/reddit-button-n357841; Alex Hern, "Reddit's Mysterious Button Experiment is Over," *Guardian,* June 8, 2015, www.theguardian.com/technology/2015/jun/08/reddits-mysterious-button-experiment-is-over; Rich McCormick, "How Reddit's Mysterious April Fools' Button Inspired Religions and Cults," The Verge, June 9, 2015, www.theverge.com/2015/6/9/8749897/reddit-april-fools-the-button-experiment-end.

126 In 1971, a: Michael D. Zeiler, "Fixed-Interval Behavior: Effects of Percentage Reinforcement," *Journal of the Experimental Analysis of Behavior* 17, no. 2 (March 1972): 177–89. See also, Michael D. Zeiler, "Fixed and Variable Schedules of Response-Independent Reinforcement," *Journal of the Experimental Analysis of Behavior* 11, no. 4 (July 1968): 405–14.

128 It's hard to exaggerate: See, for example, Jason Kincaid, "Facebook Activates 'Like' Button; Friend Feed Tires of Sincere Flattery," TechCrunch, February 9, 2009, techcrunch.com/2009/02/09/facebook-activates-like-button-friendfeed-tires-of-sincere-flattery/; M. G. Siegler, "Facebook: We'll Serve 1 Billion Likes on the Web in Just 24 Hours," TechCrunch, April 21, 2010, techcrunch.com/2010/04/21/facebook-like-button/; Erick Schonfeld, "Zuckerberg: 'We Are Building a Web Where the Default Is Social,'" TechCrunch, April 21, 2010, techcrunch.com/2010/04/21/zuckerbergs-buildin-web-default-social/.

128 The app's founder: See more on Chawla and the Lovematically app on the platform's homepage: fueled.com/lovematically/. Dozens of outlets covered Lovematically's brief rise and fall, e.g., Brendan O'Connor, "Lovematically: The Social Experiment That Instagram Shut Down after Two Hours," The Daily Dot, February 17, 2014, www.dailydot.com/technology/lovematically-auto-like-instagram-shut-down/; Jeff Bercovici, "Instagram App Lovematically Highlights, and Hijacks, the Power of the 'Like,'" *Forbes,* February 14, 2014, www.forbes.com/sites/jeffbercovici/2014/02/14/instagram-app-lovematically-highlights-and-hijacks-the-power-of-the-like/#329d9c1b64b6; Lance Ulanoff, "Why I Flooded Instagram with Likes," Mashable, February 14, 2014, mashable.com/2014/02/14/lovematically-instagram/.

130 I stumbled on: You can play Sign of the Zodiac here (but make sure you clear your schedule for several hours first): www.freeslots.co.uk/sign-of-the-zodiac/index.htm.

130 For thirteen years: Schüll's terrific book: Natasha Dow Schüll, *Addiction by Design: Machine Gambling in Las Vegas* (Princeton, NJ: Princeton University Press, 2013).

133 Mike Dixon, a: Mike Dixon and others, "Losses Disguised As Wins in Modern Multi-Line Video Slot Machines," *Addiction* 105, no. 10 (October 2010): 1819–24.

136 Bennett Foddy, who: You can find Foddy's game archive here: www.foddy.net/.

137 The game Candy: See, e.g., Joe White, "Freemium App Candy Crush Saga Earns a Record-Breaking $633,000 Each Day," AppAdvice. July 9, 2013, appadvice.com/appnn/2013/07/freemium-app-candy-crush-saga-earns-a-record-breaking-633000-each-day; Andrew Webster, "Half a Billion People Have Installed 'Candy Crush Saga,'" The Verge, November 5, 2013, www.theverge.com/2013/11/15/5107794/candy-crush-saga-500-million-downloads; Victoria Woollaston, "Candy Crush Saga Soars above Angry Birds to Become World's Most Popular Game," *Daily Mail* Online, May 14, 2013, www.dailymail.co.uk/sciencetech/article-2324228/Candy-Crush-Saga-overtakes-Angry-Birds-WORLDS-popular-game.html; Mark Walton, "Humanity Weeps As Candy Crush Saga Comes Preinstalled with Windows 10," Ars Technica. May 15, 2015, arstechnica.com/gaming/2015/05/humanity-weeps-as-candy-crush-saga-comes-pre-installed-with-windows-10/; Michael Harper, "Candy Crush Particularly Addictive—and Expensive—for Women," Redorbit, October 21, 2013, www.redorbit.com/news/technology/1112980142/candy-crush-addictive-for-women-102113/; Hayden Manders, "Candy Crush Saga Is Virtual Crack to Women," Refinery29, October 17, 2013, www.refinery29.com/2013/10/55594/candy-crush-addiction.

138　Michael Barrus and: Michael M. Barrus and Catharine A. Winstanley, "Dopamine D3 Receptors Modulate the Ability of Win-Paired Cues to Increase Risky Choice in a Rat Gambling Task," *Journal of Neuroscience* 36, no. 3 (January 2016): 785–94; K. G. Orphanides, "Scientists Built a 'Rat Casino' and It Made Rodents Riskier Gamblers," wired. co.uk, January 21, 2016, www.wired.co.uk/news/archive/2016-01/21/rat-casino-light -sound-gambling-risk; video of Barrus and Winstanley describing their results: ubbpublic- affairs, "UBC 'Rat Casino' Providing Insight into Gambling Addiction," YouTube, January 18, 2016, www.youtube.com/watch?v=6PxGnk62wGA.

139　The most powerful: On virtual reality and Oculus: Sophie Curtis, "Oculus VR: The $2bn Virtual Reality Company That Is Revolutionising Gaming," *Telegraph*, March 26, 2014, www.telegraph.co.uk/technology/video-games/video-game-news/10723562/Oculus -VR-the-2bn-virtual-reality-company-that-is-revolutionising-gaming.html; Mark Zucker- berg's Facebook announcement about the company's acquisition of Oculus VR: www .facebook.com/zuck/posts/10101319050523971; Jeff Grubb, "Oculus Founder: Rift VR Headset Is 'Fancy Wine'; Google Cardboard Is 'Muddy Water,'" VentureBeat, December 24, 2015, venturebeat.com/2015/12/24/oculus-founder-rift-vr-headset-is-fancy-wine -google-cardboard-is-muddy-water/; Stuart Dredge, "Three Really Real Questions about the Future of Virtual Reality," *Guardian*, www.theguardian.com/technology/2016/jan/07/ virtual-reality-future-oculus-rift-vr.

140　In a podcast: The Bill Simmons Podcast, "Ep. 95: Billionaire Investor Chris Sacca," The Ringer, April 28, 2016, soundcloud.com/the-bill-simmons-podcast/ep-95-billionaire -investor-chris-sacca.

144　Emily Balcetis and: Emily Balcetis, and David Dunning, "See What You Want to See: Motivational Influences on Visual Perception," *Journal of Personality and Social Psychology* 91, (2006): 612–25.

145　In a classic early: Rich Moore (director), *The Simpsons*, "Homer's Night Out," 20th Century Fox Television, Episode 10, March 25, 1990.

CHAPTER 6: PROGRESS

147　Shigeru Miyamoto knows: Background on Miyamoto and Super Mario Bros.: Wikia page for Super Mario Bros.: nintendo.wikia.com/wiki/Super_Mario_Bros.; Gus Turner, "Play- ing 'Super Mario Bros.' Can Teach You How to Design the Perfect Video Game," Com- plex, June 5, 2014, www.complex.com/Pop-Culture/2014/06/Playing-Super -Mario-Bros-Teaches-You-How-To-Design-The-Perfect-Video-Game; video explaining the features that make Super Mario Bros. so compelling: Extra Credits, "Design Club: Super Mario Bros: Level 1-1—How Super Mario Mastered Level Design," YouTube, June 5, 2014, www.youtube.com/watch?v=ZH2wGpEZVgE; NPR Staff, "Q&A: Shigeru Miyamoto on the Origins of Nintendo's Famous Characters," NPR: All Tech Considered, June 19, 2015, www.npr.org/sections/alltechconsidered/2015/06/19/415568892/q-a-shigeru -miyamoto-on-the-origins-of-nintendos-famous-characters.

149　Shubik described the: Background on Shubik's Dollar Auction Game: Martin Shubik, "The Dollar Auction Game: A Paradox in Noncooperative Behavior and Escalation," *Jour- nal of Conflict Resolution* 15, no. 1 (March 1971): 109–11.

152　You can see the same loss: Scathing Consumer Reports reviews of those sites: www.consumerreports.org/cro/2011/12/with-penny-auctions-you-can-spend-a-bundle -but-still-leave-empty-handed/index.htm.

155　when Shigeru Miyamoto: Miyamoto quote on his philosophy: Chris Johnston and Game- spot Staff, "Miyamoto Talks Dolphin at Space World," *Gamespot*, April 27, 2000, www .gamespot.com/articles/miyamoto-talks-dolphin-at-space-world-and14599/1100-2460819/.

155　"Predatory games are: Background on Adam Saltsman from an interview; also from Adam Saltsman, "Contrivance and Extortion: In-App Purchases & Microtransactions,"

Gamasutra October 18, 2011, www.gamasutra.com/blogs/AdamSaltsman/20111018/8685/ Contrivance_and_Extortion_InApp_Purchases__Microtransactions.php.

158 **This rise was:** H. Popkin, "Kim Kardashian and Her In-App Purchases Must Be Stopped!," Readwrite, July 24, 2014, readwrite.com/2014/07/24/free-mobile-games-in-app -purchases-addiction-predatory/ (page discontinued); Maya Kosoff, "Kim Kardashian's Mobile Game Won't Make Nearly As Much Money As Analysts Predicted," Business Insider, January 13, 2015, www.businessinsider.com/kim-kardashian-hollywood-mobile -game-wont-make-200-million-2015-1; Milo Yiannopoulos, "I Am Powerless to Resist the Kim Kardashian App—So I Had to Uninstall It," *Business Insider*, July 25, 2014, www.businessinsider.com/kim-kardashian-app-addicting-2014-7; Tracie Egan Morrissey, "Oh God, I Spent $494.04 Playing the Kim Kardashian Hollywood App," Jezebel, July 1, 2014, http://jezebel.com/oh-god-i-spent-494-04-playing-the-kim-kardashian-holl -1597154346.

161 **More than twenty years later:** Adam Alter, David Berri, Griffin Edwards, and Heather Kappes, "Hardship Inoculation Improves Performance but Dampens Motivation," unpub-lished manuscript (2016).

161 **Nick Yee, who:** Nick Yee completed a PhD focusing on social sciences and gaming at Stanford; he lists beginner's luck as one of the major drivers of repeat behavior in games. See: www.nickyee.com/ and www.nickyee.com/hub/addiction/attraction.html.

163 **Earlier I mentioned:** Simon Parkin, "Don't Stop: The Game That Conquered Smart-phones," *New Yorker*, June 7, 2013, www.newyorker.com/tech/elements/dont-stop-the -game-that-conquered-smartphones.

164 *Time* **called the:** Dan Fletcher, "The 50 Worst Inventions—No. 9: FarmVille," *Time*, May 27, 2010, content.time.com/time/specials/packages/article/0,28804,1991915_1991909_ 1991768,00.html.

165 **in 2010 she:** See more on Young's center here: netaddiction.com/.

CHAPTER 7: ESCALATION

168 **In the summer of 2014:** Timothy D. Wilson and others, "Just Think: The Challenges of the Disengaged Mind," *Science* 345, no. 6192 (July 2014): 75–77.

170 **In 1984, Alexey:** On Pajitnov and Tetris: Jeffrey Goldsmith, "This Is Your Brain on Tetris," *Wired*, May 1, 1994, archive.wired.com/wired/archive/2.05/tetris.html; Laurence Dodds, "The Healing Power of Tetris Has Its Dark Side," *Telegraph*, July 7, 2015, www .telegraph.co.uk/technology/video-games/11722064/The-healing-power-of-Tetris-has-its -dark-side.html; Guinness World Records, "First Videogame to Improve Brain Func-tioning and Efficiency: Tetris," n.d., www.guinnessworldrecords.com/world-records/first -video-game-to-improve-brain-functioning-and-efficiency; Richard J. Haier and others, "Regional Glucose Metabolic Changes after Learning a Complex Visuospatial/Motor Task: A Positron Emission Tomographic Study," *Brain Research* 570, nos. 1–2 (January 1992): 134–143; Mark Yates, "What Are the Benefits of Tetris?," BBC, September 3, 2009, news.bbc.co.uk/2/hi/uk_news/magazine/8233850.stm; documentary on Pajitnov and the origins of Tetris: OBZURV, "Tetris! The Story of the Most Popular Video Game," You-Tube, June 3, 2015, www.youtube.com/watch?v=8yeSnoYHmPc; Robert Stickgold and others, "Replaying the Game: Hypnagogic Images in Normals and Amnesics," *Science* 290, no. 5490 (October 2000): 350–53; Emily A. Holmes, Ella L. James, Thomas Coode-Bate, and Catherine Deeprose, "Can Playing the Computer Game 'Tetris' Reduce the Build-Up of Flashbacks for Trauma? A Proposal from Cognitive Science." *Plos One* 4, January 7, 2009e4153.

173 **In one experiment run:** Michael I. Norton, Daniel Mochon, and Dan Ariely, "The 'IKEA Effect': When Labor Leads to Love," *Journal of Consumer Psychology* 22, no. 3 (July 2012): 453–60; see also: Dan Ariely, Emir Kamenica, and Dražen Prelec, "Man's Search for

Meaning: The Case of Legos," *Journal of Economic Behavior and Organization* 67 (2008): 671–77.

174 **Vygotsky explained that:** On Vygotsky and Csikszentmihalyi: L. S. Vygotsky, *Mind in Society: Development of Higher Psychological Processes* (Cambridge, MA: Harvard University Press, 1978); Mihaly Csikszentmihalyi, *Flow: The Psychology of Optimal Experience* (New York: Harper & Row, 1990); Fausto Massimini, Mihaly Csikszentmihalyi, and Massimo Carli, "The Monitoring of Optimal Experience: A Tool for Psychiatric Rehabilitation," *Journal of Nervous and Mental Disease* 175, no. 9 (September 1987): 545–9.

178 **I remember playing a:** IGN Staff, "PC Retroview: Myst," *IGN*, August 1, 2000, www.ign.com/articles/2000/08/01/pc-retroview-myst.

179 **an Irish game:** Information on this section comes from an interview with Bennett Foddy, and from the following references: J. C. Fletcher, "Terry Cavanagh Goes Inside Super Hexagon," Engadget, September 9, 2012, www.engadget.com/2012/09/21/terry-cavanagh-goes-inside-super-hexagon; video of Terry Cavanagh completing the impossibly quick final level of Super Hexagon at a gaming conference: Fantastic Arcade, "Terry Cavanagh Completes Hyper Hexagonest Mode in Super Hexagon on Stage (78:32)," YouTube, September 21, 2012, www.youtube.com/watch?v=JJ96olZr8DE.

181 **We know this from a paper:** In 2015, two marketing professors published a paper about near wins: Monica Wadhwa, and JeeHye Christine Kim, "Can a Near Win Kindle Motivation? The Impact of Nearly Winning on Motivation for Unrelated Rewards," *Psychological Science* 26 (2015): 701–8; see also: Győző Kurucz and Attila Körmendi, "Can We Perceive Near Miss? An Empirical Study," *Journal of Gambling Studies* 28, no. 1 (February 2011): 105–11.

183 **Neither one signals:** Note that it is legal to change how losses are presented, so a near win is just as legal as a clear loss.

183 **During the 1990s:** See: Paco Underhill, *Why We Buy: The Science of Shopping* (New York: Simon and Schuster, 1999).

185 **I've never used:** See, e.g., J. Etkin, "The Hidden Cost of Personal Quantification," *Journal of Consumer Research*, forthcoming).

186 **The same technology:** On overworking and karoshi, see: Daniel S. Hamermesh, and Elena Stancanelli, "Long Workweeks and Strange Hours," *Industrial and Labor Relations Review* (forthcoming); Christopher K. Hsee, Jiao Zhang, Cindy F. Cai, and Shirley Zhang, "Overearning," *Psychological Science* 24 (2013): 852–59; Lauren F. Friedman, "Here's Why People Work Like Crazy, Even When They Have Everything They Need," Business Insider, July 10, 2014, www.businessinsider.com/why-people-work-too-much-2014-7; International Labour Organization, "Case Study: Karoshi: Death from Overwork," *International Labour Relations*, April 23, 2013, www.ilo.org/safework/info/publications/WCMS_211571/lang—en/index.htm ; China Post News Staff, "Overwork Confirmed to Be Cause of Nanya Engineer's Death," *China Post*, October 15, 2011, www.chinapost.com.tw/taiwan/national/national-news/2011/03/15/294686/Overwork-confirmed.htm.

188 **In a classic paper:** Dražen Prelec and Duncan Simester, "Always Leave Home Without It: A Further Investigation of the Credit-Card Effect on Willingness to Pay, *Marketing Letters* 12, no. 1 (2001): 5–12; see also: Dražen Prelec and George Loewenstein, "The Red and the Black: Mental Accounting of Savings and Debt," *Marketing Science* 17, no. 1 (1998): 4–28.

CHAPTER 8: CLIFFHANGERS

191 **In their own:** Responses to the ending of *The Italian Job* on the Internet Movie Database: www.imdb.com/title/tt0064505/reviews.

192 **Forty years earlier:** Background material on Bluma Zeigarnik and her eponymous effect: A. V. Zeigarnik, "Bluma Zeigarnik: A Memoir," *Gestalt Theory* 29, no. 3 (December 8, 2007): 256–68; Bluma Zeigarnik, "On Finished and Unfinished Tasks," in *A Source Book* *of Gestalt Psychology*, W. D. Ellis, ed., (New York: Harcourt, Brace, and Company, 1938), 300–14; Colleen M. Seifert, and Andrea L. Patalano, "Memory for Incomplete Tasks: A Re-Examination of the Zeigarnik Effect," in *Proceedings of the Thirteenth Annual Conference of the Cognitive Science Society* (Mahwah, NJ: Erlbaum, 1991), 114–19.

194 **"September" is a:** Dan Charnas, "The Song That Never Ends: Why Earth, Wind & Fire's 'September' Sustains," NPR, September 19, 2014, www.npr.org/2014/09/19/349621429/the-song-that-never-ends-why-earth-wind-fires-september-sustains; interview with Verdine White about the melody and popularity of "September" at Songfacts: www.songfacts.com/blog/interviews/verdine_white_of_earth_wind_fire/.

196 **In October 2014:** On *Serial* and *Making a Murderer*: Louise Kiernan, "'Serial' Podcast Producers Talk Storytelling, Structure and If They Know Whodunnit," Nieman Storyboard, October 30, 2014, http://niemanstoryboard.org/stories/serial-podcast-producers-talk-story telling-structure-and-if-they-know-whodunnit/; Jeff Labrecque, "'Serial' Podcast Makes Thursdays a Must-Listen Event," *Entertainment Weekly*, October 30, 2014, www.ew.com/article/2014/10/30/serial-podcast-thursdays; Josephine Yurcaba, "This American Crime: Sarah Koenig on Her Hit Podcast 'Serial,'" *Rolling Stone*, October 24, 2014, www.rolling stone.com/culture/features/sarah-koenig-on-serial-20141024; Maria Elena Fernandez, "'Serial': The Highly Addictive Spinoff Podcast of 'This American Life,'" NBC News, October 30, 2014, www.nbcnews.com/pop-culture/viral/serial-highly-addictive-spinoff-podcast-american-life-n235751; John Boone, "The 13 Stages of Being Addicted to 'Serial,'" ET Online, November 12, 2014, www.etonline.com/news/153862_the_13_stages_of_being_addicted_to_serial/; Yoni Heisler, "'Making a Murderer' Is the Most Addictive Show Netflix Has Ever Released," Yahoo Tech, January 14, 2016, www.yahoo.com/tech/making-murderer-most-addictive-show-netflix-ever-released-143343536.html.

201 **When David Chase:** James Greenberg, "This Magic Moment," Directors Guild of America, Spring 2015, www.dga.org/Craft/DGAQ/All-Articles/1502-Spring-2015/Shot-to-Remem ber-The-Sopranos.aspx; Alan Sepinwall, "David Chase Speaks!," NJ.com, June 11, 2007, blog.nj.com/alltv/2007/06/david_chase_speaks.html; Maureen Ryan, "Are You Kidding Me? That Was the Ending of 'The Sopranos'?," *Chicago Tribune*, June 10, 2007, featuresblogs .chicagotribune.com/entertainment_tv/2007/06/are_you_kidding.html.

204 **But, in 2001:** Gregory S. Berns, Samuel M. McClure, Giuseppe Pagnoni, and P. Read Montague, "Predictability Modulates Human Brain Response to Reward," *Journal of Neuroscience* 21, no. 8 (April 2001): 2793–98. See also: Gregory S. Berns, *Satisfaction: The Science of Finding True Fulfillment* (New York: Henry Holt & Co., 2005).

206 **Darleen Meier, who:** Tara Parker-Pope, "This Is Your Brain at the Mall: Why Shopping Makes You Feel So Good," *Wall Street Journal*, December 6, 2005, online.wsj.com/ad/article/cigna/SB113382650575214543.html; Amanda M. Fairbanks, "Gilt Addicts Anonymous: The Daily Online Flash Sale Fixation, Huffington Post, December 22, 2011, www .huffingtonpost.com/2011/12/22/gilt-shopping-addiction_n_1164035.html; Elaheh Nozari, "Inside the Facebook Group for People Addicted to QVC," The Kernel, January 31, 2016, kernelmag.dailydot.com/issue-sections/headline-story/15703/qvc-shopping-addiction-face book-group/; Darleen Meier's blog entries: darlingdarleen.com/2010/12/gilt-addic/, darlingdarleen.com/2010/10/gi/; message board posts by Cassandra, another Gilt addict: forum.purseblog.com/general-shopping/woes-of-a-gilt-addict-should-i-ban-658398.html.

209 **Psychologists Eric Johnson:** Eric J. Johnson and Daniel Goldstein, "Do Defaults Save Lives?," *Science*, 302, no. 5649 (November 2003): 1338–39.

211 **Search term popularity:** Netflix research on binge-watching: Kelly West, "Unsurprising: Netflix Survey Indicates People Like to Binge-Watch TV," CinemaBlend, 2014, www .cinemablend.com/television/Unsurprising-Netflix-Survey-Indicates-People-Like-Binge -Watch-TV-61045.html.

211 **Netflix found similar:** John Koblin. "Netflix Studied Your Binge-watching Habit. That

Didn't Take Long," *New York Times*, June 8, 2016, www.nytimes.com/2016/06/09/business/media/netflix-studied-your-binge-watching-habit-it-didnt-take-long.html; "Netflix & Binge: New Binge Scale Reveals TV Series We Devour and Those We Savor," Netflix, June 8, 2016. media.netflix.com/en/press-releases/netflix-binge-new-binge-scale-reveals-tv-series-we-devour-and-those-we-savor-1.

CHAPTER 9: SOCIAL INTERACTION

214 In December 2009: On the divergent fortunes of Instagram and Hipstamatic: Shane Richmond, "Instagram, Hipstamatic, and the Mobile Technology Movement," *Telegraph*, August 19, 2011, www.telegraph.co.uk/technology/news/8710979/Instagram-Hipstamatic-and-the-mobile-photography-movement.html; Marty Yawnick, "Q&A: Hipstamatic: The Story Behind the Plastic App with the Golden Shutter," Life in Lofi, January 7, 2010, lifeinlofi.com/2010/01/07/qa-hipstamatic-the-story-behind-the-plastic-app-with-the-golden-shutter/; Marty Yawnick, "News: Wausau City Pages Uncovers the Real Hipstamatic Backstory?," Life in Lofi, December 23, 2010, lifeinlofi.com/2010/12/23/news-wausau-city-pages-uncovers-the-real-hipstamatic-backstory/; the (arguably fabricated) "history" of Hipstamatic and original Hipstamatic 100 camera: history.hipstamatic.com/; Libby Plummer, "Hipstamatic: Behind the Lens," Pocket-lint, November 16, 2010, www.pocket-lint.com/news/106994-hipstamatic-iphone-app-android-interview. Damon Winter's photos that contributed to Instagram's early rise: James Estrin, "Finding the Right Tool to Tell a War Story," *New York Times*, November 21, 2010, lens.blogs.nytimes.com/2010/11/21/finding-the-right-tool-to-tell-a-war-story/; Katherine Rushton, "Who's Getting Rich from Facebook's $1bn Instagram deal?," *Telegraph*, April 10, 2012, www.telegraph.co.uk/technology/facebook/9195380/Whos-getting-rich-from-Facebooks-1bn-Instagram-deal.html; an excellent article on how Facebook's purchase of Instagram affected Hipstamatic's dejected founders: Nicole Carter and Andrew MacLean, "The Photo App Facebook Didn't Buy: Hipstamatic," Inc.com, April 12, 2012, www.inc.com/nicole-carter-and-andrew-maclean/photo-app-facebook-didnt-buy-hipstamatic.html; Joanna Stern, "Facebook Buys Instagram for $1 Billion," ABCNews.com, April 9, 2012, abcnews.go.com/blogs/technology/2012/04/facebook-buys-instagram-for-1-billion/.

218 When students at: David Dunning, *Self-Insight: Roadblocks and Detours on the Path to Knowing Thyself* (New York: Psychology Press, 2005); David Dunning, Judith A. Meyerowitz, and Amy D. Holzberg, "Ambiguity and Self-Evaluation: The Role of Idiosyncratic Trait Definitions in Self-Serving Assessments of Ability," *Journal of Personality and Social Psychology* 57, no. 6 (December 1989): 1082–90.

219 Psychologists call this the: Roy F. Baumeister, Ellen Bratslavsky, Catrin Finkenauer, and Kathleen D. Vohs, "Bad Is Stronger Than Good," *Review of General Psychology* 5, no. 4 (2001): 323–70; Mark D. Pagel, William W. Erdly, and Joseph Becker, "Social Networks: We Get By with (and in Spite of) a Little Help from Our Friends," *Journal of Personality and Social Psychology* 53, no. 4 (October 1987): 793–804; John F. Finch and others, "Positive and Negative Social Ties among Older Adults: Measurement Models and the Prediction of Psychological Distress and Well-Being," *American Journal of Community Psychology* 17, no. 5 (October 1989): 585–605; Brenda Major and others, "Mixed Messages: Implications of Social Conflict and Social Support Within Close Relationships for Adjustment to a Stressful Life Event," *Journal of Personality and Social Psychology* 72, no. 6 (June 1997): 1349–63; Amiram D. Vinokur and Michelle van Ryn, "Social Support and Undermining in Close Relationships: Their Independent Effects on the Mental Health of Unemployed Persons," *Journal of Personality and Social Psychology* 65, no. 2 (1993): 350–59; Hans Kreitler and Shulamith Kreitler, "Unhappy Memories of the 'Happy Past': Studies in Cognitive Dissonance," *British Journal of Psychology* 59, no. 2 (May 1968): 157–66; Mark R. Leary, Ellen S. Tambor, Sonja K. Terdal, and Deborah L. Downs, "Self-Esteem As an Interpersonal

Monitor: The Sociometer Hypothesis," *Journal of Personality and Social Psychology* 68, no. 3 (1995): 518–30.

220 **Essena O'Neill, a:** Elle Hunt, "Essena O'Neill Quits Instagram Claiming Social Media 'Is Not Real Life,'" *Guardian*, November 3, 2015, www.theguardian.com/media/2015/nov/03/instagram-star-essena-oneill-quits-2d-life-to-reveal-true-story-behind-images; Megan Mc-Cluskey, "Instagram Star Essena O'Neill Breaks Her Silence on Quitting Social Media," *Time*, January 5, 2015, time.com/4167856/essena-oneill-breaks-silence-on-quitting-social -media/; O'Neill describes her perspective in this video: Essena O'Neill, "Essena O'Neill— Why I REALLY Am Quitting Social Media," YouTube, November 3, 2015, www.youtube .com/watch?v=gmAbwTQvWX8.

222 **The site, which:** On Hot or Not and its founders: Alexia Tsotsis, "Facemash.com, Home of Zuckerberg's Facebook Predecessor, for Sale," TechCrunch, October 5, 2010, tech crunch.com/2010/10/05/facemash-sale/; Alan Farnham, "Hot or Not's Co-Founders: Where Are They Now?," ABCNews.com, June 2, 2014, abcnews.go.com/Business/found ers-hot-today/story?id=23901082; David Pescovitz, "Cool Alumni: HOTorNOT.com Founders James Hong and Jim Young," *Lab Notes*, October 1, 2004, coe.berkeley.edu/labnotes/1004/coolalum.html; Liz Gannes, "Hot or Not Creator James Hong Doesn't Care If He Strikes It Rich or Not with New App," Recode.net, November 21, 2014. recode .net/2014/11/21/james-hong-doesnt-want-to-be-a-billionaire-but-he-does-want-you -to-think-hes-relevant/.

225 **One user who:** Manitou2121 appended the following note below his Hot or Not composite images: "These women do not exist. They each are a composite of about thirty faces that I created to find out the current standard of good looks on the Internet. On the popular Hot or Not website, people rate others' attractiveness on a scale of 1 to 10. An average score based on hundreds or even thousands of individual ratings takes only a few days to emerge. I collected some photos from the site, sorted them by rank and used SquirlzMorph to create multimorph composites from them. Unlike projects like Face of Tomorrow or Beauty Check where the subjects are posed for the purpose, the portraits are blurry because the source images are low resolution with differences in posture, hair styles, glasses, etc., so that I could use only thiry-six control points for the morphs. What did I conclude about good looks from these virtual faces? First, morphs tend to be prettier than their sources because face asymmetries and skin blemishes average out. However, the low score images show that fat is not attractive. The high scores tend to have narrow faces. I will leave it to you to find more differences and to do a similar project for men." commons.wikimedia.org/wiki/File: Hotornot_comparisons_manitou2121.jpg.

226 **Psychologists call this perfect:** Marilynn B. Brewer, "The Social Self: On Being the Same and Different at the Same Time," *Personality and Social Psychology Bulletin* 17, no. 5 (October 1991): 475–82; Marilynn B. Brewer and Sonia Roccas, "Individual Values, Social Identity, and Optimal Distinctiveness," in *Individual Self, Relative Self, Collective Self*, C. Sedikides & M. Brewer, eds. (Philadelphia, PA: Psychology Press, 2001), 219–37.

228 **Hilarie Cash, a:** Many of Cash's ideas on the importance of face-to-face interactions are reflected in: Thomas Lewis, Fari Amini, and Richard Lannon, *A General Theory of Love* (New York: Random House, 2001).

230 **Cash suggested I:** Background on Andy Doan's ideas and amblyopia: Andrew K. Przybylski, "Electronic Gaming and Psychosocial Adjustment," *Pediatrics*, 134, (2014): e716-e722; Colin Blakemore, and Grahame F. Cooper, "Development of the Brain Depends on the Visual Environment," *Nature* 228 (October 1970): 477–78; Wilder Penfield and Lamar Roberts, *Speech and Brain-Mechanisms* (Princeton, NJ: Princeton University Press, 1959).

233 **One study found:** Details on the study are available at the iKeepSafe website: ikeep safe.org/be-a-pro/balance/too-much-time-online/.

CHAPTER 10: NIPPING ADDICTIONS AT BIRTH

237 In the summer of 2012: Experiment and background literature on the summer of 2012: Yalda T. Uhls and others, "Five Days at Outdoor Education Camp Without Screens Improves Preteen Skills with Nonverbal Emotion Cues," *Computers in Human Behavior* 39 (October 2014): 387–92; Sandra L. Hofferth, "Home Media and Children's Achievement and Behavior," *Child Development* 81, no. 5 (September–October 2010): 1598–1619; Internet World Stats: www.Internetworldstats.com/stats.htm; Victoria J. Rideout, Ulla G. Foehr, and Donald F. Roberts, *Generation M2: Media in the Lives of 8- to 18-Year-Olds* (Menlo Park: CA: Kaiser Family Foundation, 2010); Amanda Lenhart, *Teens, Smartphones & Texting* (Washingon, DC: Pew Research Center, 2010); Jay N. Giedd, "The Digital Revolution and Adolescent Brain Evolution," *Journal of Adolescent Health* 51, no. 2 (August 2012): 101–5; Stephen Nowicki and John Carton, "The Measurement of Emotional Intensity from Facial Expressions," *Journal of Social Psychology* 133, no. 5 (November 1993): 749–50; Stephen Nowicki, *Manual for the Receptive Tests of the DANVA2*. To find sample items from the DANVA test, including the adult test, see: psychology.emory.edu/labs/interper sonal/Adult/danva.swf.

240 Why shouldn't kids: In preparing this chapter, I read dozens of reports on screen exposure among kids. They explored not only whether kids should be exposed to screens, but when exposure should begin, how much was okay, and how screens should be introduced. These reports and references include: Claire Lerner and Rachel Barr, "Screen Sense: Setting the Record Straight," 2014, www.zerotothree.org/parenting-resources/screen-sense-setting-the-record-straight; in particular, see this exchange at the Huffington Post, which consisted of one column decrying screens, and two replies that challenged and clarified the original column: Cris Rowan, "10 Reasons Why Handheld Devices Should Be Banned for Children under 12," Huffington Post, March 6, 2014, m.huffpost.com/us/entry/10-reasons-why-handheld-devices-should-be-banned_b_4899218.html, David Kleeman, "10 Reasons Why We Need Research Literacy, Not Scare Columns," Huffington Post, March 11, 2014, www. huffingtonpost.com/david-kleeman/10-reasons-why-we-need-re_b_4940987.html,   Lisa Nielsen, "10 Points Where the Research Behind Banning Handheld Devices in Children Is Flawed," Huffington Post, March 24, 2014, www.huffingtonpost.com/lisa-nielsen/10-reasons-why-the-resear_b_5004413.html?1395687657; UserExperiencesWorks, "A Magazine Is an iPad That Does Not Work," YouTube, October 6, 2011, www.youtube.com/watch?v=aXV-yaFmQNk; American Academy of Pediatrics, "Media and Children," 2015, www.aap.org/en-us/advocacy-and-policy/aap-health-initiatives/pages/media-and-children .aspx; Lisa Guernsey, "Common-Sense, Science-Based Advice on Toddler Screen Time," Slate, November 13, 2014, www.slate.com/articles/technology/future_tense/2014/11/zero_ to_three_issues_common_sense_advice_on_toddler_screen_time.html; Farhad Manjoo, "Go Ahead, a Little TV Won't Hurt Him," Slate, October 12, 2011, www.slate.com/articles/technology/technology/2011/10/how_much_tv_should_kids_watch_why_doctors_prohibitions_on_screen.html; Kaiser Foundation, "The Media Family: Electronic Media in the Lives of Infants, Toddlers, Preschoolers, and Their Parents," 2006, kaiser-familyfoundation.files.wordpress.com/2013/01/7500.pdf; Erika Hoff, "How Social Contexts Support and Shape Language Development," *Developmental Review* 26, no. 1 (March 2006): 55–88; Nancy Darling and Laurence Steinberg, "Parenting Style As Context: An Integrative Model," *Psychological Bulletin* 113, no. 3 (1993): 487–96; Annie Bernier, Stephanie M. Carlson, and Natasha Whipple, "From External Regulation to Self-Regulation: Early Parenting Precursors of Young Children's Executive Functioning," *Child Development* 81, no. 1 (January 2010): 326–39; Susan H. Landry, Karen E. Smith, and Paul R. Swank, "The Importance of Parenting During Early Childhood for School-Age Development," *Developmental Neuropsychology* 24, nos. 2–3 (2003): 559–91; Sarah Roseberry, Kathy Hirsh-Pasek, and Roberta M. Golinkoff, "Skype Me! Socially Contingent Interactions

Help Toddlers Learn Language," *Child Development* 85, no. 3 (May–June 2014): 956–70; Angeline S. Lillard and Jennifer Peterson, "The Immediate Impact of Different Types of Television on Young Children's Executive Function," *Pediatrics* 128, No. 4 (October 2011): 644–49; N. Brito, R. Barr, P. McIntyre, and G. Simcock, "Long-Term Transfer of Learning from Books and Video During Toddlerhood," *Journal of Experimental Child Psychology* 111, no. 1 (January 2012): 108–19; Rachel Barr and Harlene Hayne, "Developmental Changes in Imitation from Television During Infancy," *Child Development* 70, no. 5 (September–October 1999): 1067–81; Jane E. Brody, "Screen Addiction Is Taking a Toll on Children," *New York Times*, July 6, 2015, well.blogs.nytimes.com/2015/07/06/screen-addiction-is-taking-a-toll-on-children/; Conor Dougherty, "Addicted to Your Phone? There's Help for That," *New York Times*, July 11, 2015, www.nytimes.com/2015/07/12/sunday-review/addicted-to-your-phone-theres-help-for-that.html; Alejandrina Cristia and Amanda Seidl, "Parental Reports on Touch Screen Use in Early Childhood," Plos One 10(6) (2015): e0128338, doi:10.1371/journal.pone.0128338; C. S. Green and D. Bavelier, "Exercising Your Brain: A Review of Human Brain Plasticity and Training-Induced Learning," *Psychology and Aging* 23, no. 4 (December 2008): 692–701; Kathy Hirsh-Pasek and others, "Putting Education in 'Educational' Apps: Lessons from the Science of Learning, *Psychological Science in the Public Interest* 16, no. 1 (2015): 3–34; Deborah L. Linebarger, Rachel Barr, Matthew A. Lapierre, and Jessica T. Piotrowski, "Associations Between Parenting, Media Use, Cumulative Risk, and Children's Executive Functioning," *Journal of Developmental & Behavioral Pediatrics* 35, no. 6 (July–August 2014): 367–77; Jessi Hempel, "How about a Social Media Sabbatical? *Wired* Readers Weigh In," *Wired*, August 5, 2015, www.wired.com/2015/08/social-media-sabbatical-wired-readers-weigh/; "'Digital Amnesia' Leaves Us Vulnerable, Survey Suggests," CBC News, October 8, 2015, www.cbc.ca/news/technology/digital-amnesia-kaspersky-1.3262600 (link to the report available in the body of the article.)

241 (David Denby, a: David Denby, "Do Teens Read Seriously Anymore?," *New Yorker*, February 23, 2016, www.newyorker.com/culture/cultural-comment/books-smell-like-old-people-the-decline-of-teen-reading.

242 The M.I.T. psychologist: See: Sherry Turkle, *Reclaiming Conversation: The Power of Talk in a Digital Age* (New York: Penguin Press, 2015); Sherry Turkle, *Alone Together: Why We Expect More from Technology and Less from Each Other* (New York: Basic Books, 2011).

250 Catherine Steiner-Adair, the psychologist: Catherine Steiner-Adair, *The Big Disconnect: Protecting Childhood and Family Relationships in the Digital Age* (New York: Harper, 2013).

251 The East Asian: On the Chinese and Korean approaches to Internet addiction treatment: Shosh Shlam and Hilla Medalia, *Web Junkie*, 2013; see also: Whitney Mallett, "Behind 'Web Junkie,' a Documentary about China's Internet-Addicted Teens," *Motherboard*, January 27, 2014, motherboard.vice.com/blog/behind-web-junkie-a-documentary-about-chinas-Internet-addicted-teens.

255 Kimberly Young, the: On Kimberly Young and her Internet Addiction Test: Test available at netaddiction.com/Internet-addiction-test; Kimberly S. Young, *Caught in the Net: How to Recognize Signs of Internet Addiction—and a Winning Strategy for Recovery* (John Wiley & Sons: New York, 1998); Kimberly S. Young, "Internet Addiction: The Emergence of a New Clinical Disorder," *CyberPsychology & Behavior* 1, no. 3 (1998): 237–44; Laura Widyanto and Mary McMurran, "The Psychometric Properties of the Internet Addiction Test," *CyberPsychology & Behavior* 7, no. 4 (2004): 443–50; Man Kit Chang and Sally Pui Man Law, "Factor Structure for Young's Internet Addiction Test: A Confirmatory Study," *Computers in Human Behavior* 24, no. 6 (September 2008): 2597–2619; Yasser Khazaal and others, "French Validation of the Internet Addiction Test," *CyberPsychology & Behavior* 11, no. 6 (November 2008): 703–6; Steven Sek-yum Ngai, "Exploring the Validity of the Internet Addiction Test for Students in Grades 5–9 in Hong Kong," *International Journal of Adoles-*

cence and Youth 13, no. 3 (January 2007): 221–37; Kimberly S. Young, "Treatment Outcomes Using CBT-IA with Internet-Addicted Patients," *Journal of Behavioral Addictions* 2, no. 4 (December 2013): 209–15.

258 Carrie Wilkens, cofounder: On motivational interviewing and Carrie Wilkens: Gabrielle Glaser, "A Different Path to Fighting Addiction," *New York Times*, July 3, 2014, www .nytimes.com/2014/07/06/nyregion/a-different-path-to-fighting-addiction.html; William R. Miller and Stephen Rollnick, *Motivational Interviewing: Helping People Change,* 3rd ed., (New York: Guilford Press, 2012); William R. Miller and Paula L. Wilbourne, "Mesa Grande: A Methodological Analysis of Clinical Trials of Treatments for Alcohol Use Disorders," *Addiction* 97, no. 3 (March 2002): 265–77; Tracy O'Leary Tevyaw and Peter M. Monti, "Motivational Enhancement and Other Brief Interventions for Adolescent Substance Abuse: Foundations, Applications and Evaluations," *Addiction* 99 (December 2004): 63–75; C. Dunn, L. Deroo, and F. P. Rivara, "The Use of Brief Interventions Adapted from Motivational Interviewing Across Behavioral Domains: A Systematic Review," *Addiction* 96, no. 12 (December 2001): 1725–42; Craig S. Schwalbe, Hans Y. Oh, and Allen Zweben, "Sustaining Motivational Interviewing: A Meta-Analysis of Training Studies," *Addiction* 109, 1287–94; Kate Hall and others, "After 30 Years of Dissemination, Have We Achieved Sustained Practice Change in Motivational Interviewing?," *Addiction* (in press; a sample script is available here: careacttarget.org/sites/default/files/file-upload/resources/module5 -handout1.pdf).

260 The technique's effectiveness: Edward L. Deci and Richard M. Ryan, eds., *Handbook of Self-Determination Research* (Rochester, NY: University of Rochester Press, 2002); Mark R. Lepper, David Greene, and Richard E. Nisbett, "Undermining Children's Intrinsic Interest with Extrinsic Reward: A Test of the 'Overjustification' Hypothesis," *Journal of Personality and Social Psychology* 28 (1973): 129–37; Edward L. Deci, "Effects of Externally Mediated Rewards on Intrinsic Motivation," *Journal of Personality and Social Psychology* 18, no. 1 (April 1871): 105–15; Richard M. Ryan, "Psychological Needs and the Facilitation of Integrative Processes, *Journal of Personality* 63, no. 3 (September 1995): 397–427; Edward L. Deci, E. and Richard M. Ryan, "A Motivational Approach to Self: Integration in Personality," in *Nebraska Symposium on Motivation: Vol. 38. Perspectives on Motivation*, Richard A. Dienstbier, ed., (Lincoln, NE: University of Nebraska Press, 1991), 237–88; Edward L. Deci, and Richard M. Ryan, "Human Autonomy: The Basis for True Self-Esteem," in *Efficacy, Agency, and Self-Esteem*, Michael H. Kernis, ed., (New York: Springer, 1995); Roy F. Baumeister and Mark R. Leary, "The Need to Belong: Desire for Interpersonal Attachments As a Fundamental Human Motivation," *Psychological Bulletin* 117, no. 3 (May 1995): 497–529.

CHAPTER 11: HABITS AND ARCHITECTURE

263 In the United States, politics: Joseph M. Strayhorn and Jillian C. Strayhorn, "Religiosity and Teen Birth Rate in the United States," *Reproductive Health* 6, no. 14 (September 2009): 1–7; Benjamin Edelman, "Red Light States: Who Buys Online Adult Entertainment?," *Journal of Economic Perspectives* 23, no. 1 (Winter 2009): 209–20; Anna Freud, *The Ego and the Mechanisms of Defense* (New York: Hogarth, 1936); Cara C. MacInnis and Gordon Hodson, "Do American States with More Religious or Conservative Populations Search More for Sexual Content on Google?," *Archives of Sexual Behavior* 44 (2015): 137–47.

264 Feshbach and Singer: Much of the authors' relevant research is contained in this book: Seymour Feshbach and Robert D. Singer, *Television and Aggression: An Experimental Field Study* (San Franciso: Jossey-Bass, 1971).

265 According to Wendy: Alina Tugend, "Turning a New Year's Resolution into Action with the Facts," *New York Times*, January 9, 2015, www.nytimes.com/2015/01/10/your-money/some-facts-to-turn-your-new-years-resolutions-into-action.html.

266 **Xianchi Dai and Ayelet Fishbach at:** Xianchi Dai and Ayelet Fishbach, "How Nonconsumption Shapes Desire," *Journal of Consumer Research* 41 (December 2014): 936–52.

266 **A psychologist named Dan:** Daniel M. Wegner, "Ironic Processes of Mental Control," *Psychological Review* 101, no. 1 (1994): 34–52; Daniel M. Wegner and David J. Schneider, "The White Bear Story," *Psychological Inquiry* 14, nos. 3–4 (2003): 326–29; Daniel M. Wegner, *White Bears and Other Unwanted Thoughts: Suppression, Obsession, and the Psychology of Mental Control* (New York: Viking, 1989); Daniel M. Wegner, David J. Schneider, Samuel R. Carter III, and Teri L. White, "Paradoxical Effects of Thought Suppression," *Journal of Personality and Social Psychology* 53, no. 1 (1987): 5–13.

267 **The key to overcoming:** On changing habits by replacement and distraction: Christos Kouimtsidis and others, *Cognitive-Behavioural Therapy in the Treatment of Addiction* (Chichester, UK: John Wiley & Sons, 2007); Charles Duhigg, "The Golden Rule of Habit Change," PsychCentral, n.d., psychcentral.com/blog/archives/2012/07/17/the-golden-rule-of-habit-change; Charles Duhigg, *The Power of Habit: Why We Do What We Do in Life and Business* (New York: Random House, 2012); Melissa Dahl, "What If You Could Just 'Forget' to Bite Your Nails?," *New York*, July 16, 2014, nymag.com/scienceofus/2014/07/what-if-you-could-forget-to-bite-your-nails.html.

269 **An innovation agency:** Background on the Realism device: www.realismsmartdevice.com/meet-realism; "Realism: An Alternative to Our Addiction to Smartphones," Untitled Magazine, December 18, 2014, untitled-magazine.com/realism-an-alternative-to-our-addiction-to-smartphones/#.VorirVLqWPv.

270 **The key is to work:** On the importance of understanding genuine motivation: Paul Simpson, *Assessing and Treating Compulsive Internet Use* (Brentwood, TN: Cross Country Education, 2013); Kimberly Young and Cristiano Nabuco de Abreu, eds., *Internet Addiction: A Handbook and Guide to Evaluation and Treatment* (Hoboken, NJ: John Wiley & Sons, 2011).

271 **Three quarters stick:** On New Year's resolution statistics, and habit formation and duration: www.statisticbrain.com/new-years-resolution-statistics; John C. Norcross, Marci S. Mrykalo, and Matthew D. Blagys, "*Auld Lang Syne*: Success Predictors, Change Processes, and Self-Reported Outcomes of New Year's Resolvers and Nonresolvers," *Journal of Clinical Psychology* 58, no. 4 (April 2002): 397–405; Jeremy Dean, *Making Habits, Breaking Habits: Why We Do Things, Why We Don't, and How to Make Any Change Stick* (Cambridge, MA: Da Capo Press, 2013); Phillippa Lally, Cornelia H. M. van Jaarsveld, Henry W. W. Potts, and Jane Wardle, "How Are Habits Formed: Modelling Habit Formation in the Real World," *European Journal of Social Psychology* 40, no. 6 (October 2010): 998–1009.

272 **We know this works:** Vanessa M. Patrick and Henrik Hagtvedt, "'I Don't' versus 'I Can't': When Empowered Refusal Motivates Goal-Directed Behavior," *Journal of Consumer Research* 39 (2011), 371–81.

273 **That's the idea behind the:** The term "behavioral architecture" is from: Richard H. Thaler and Cass R. Sunstein, *Nudge: Improving Decisions about Health, Wealth, and Happiness* (New Haven, CT: Yale University Press, 2008).

275 **When World War II:** This section contains excerpts from a piece I wrote for 99u: Adam L. Alter, "How to Build a Collaborative Office Space Like Pixar and Google," n.d., 99u.com/articles/16408/how-to-build-a-collaborative-office-space-like-pixar-and-google; Leon Festinger, Kurt W. Back, and Stanley Schacter, *Social Pressures in Informal Groups: A Study of Human Factors in Housing* (Stanford, CA: Stanford University Press, 1950).

278 **Rewards are a:** On the power of loss aversion and motivation: Thomas C. Schelling, "Self-

Command in Practice, in Policy, and in a Theory of Rational Choice, *American Economic Review* 74, no. 2 (1984): 1–11; Jan Kubanek, Lawrence H. Snyder, and Richard A. Abrams, "Reward and Punishment Act as Distinct Factors in Guiding Behavior," *Cognition* 139 (June 2015): 154–67; Ronald G. Fryer, Steven D. Levitt, John List, and Sally Sadoff, "Enhancing the Efficacy of Teacher Incentives Through Loss Aversion: A Field Experiment," Working Paper 18237, National Bureau of Economic Research, Cambridge, MA, 2012; Daniel Kahneman and Amos Tversky, "Prospect Theory: An Analysis of Decision under Risk," *Econometrica* 47, no. 2 (March 1979): 263–92. Don't Waste Your Money game: Paul Simpson, *Assessing and Treating Compulsive Internet Use* (Brentwood, TN: Cross Country Education, 2013). Relational spending: Elizabeth Dunn and Michael Norton, *Happy Money: The Science of Happier Spending* (New York: Simon & Schuster, 2013).

285 **Benjamin Grosser, a:** The Facebook Demetricator site: bengrosser.com/projects/facebook-demetricator/.

287 **If you understand:** On binge-watching and overcoming the hook of the cliffhanger: Patrick Allan, "Overcome TV Show Binge-Watching with a Lesson in Plot," Lifehacker, September, 29, 2014, lifehacker.com/overcome-tv-show-binge-watching-with-a-lesson-in-plot -1640472646; see also: Michael Hsu, "How to Overcome a Binge-Watching Addiction," *Wall Street Journal*, September 26, 2014, www.wsj.com/articles/how-to-overcome-a-binge -watching-addiction-1411748602; this cliffhanger short-circuiting idea was originally inspired by Tom Meyvis, a colleague of mine at NYU, and Uri Simonsohn, a professor at the University of Pennsylvania's Wharton School of Business.

291 **For each show:** Jacob Kastrenakes, "Netflix Knows the Exact Episode of a TV Show That Gets You Hooked," The Verge, September 23, 2015, www.theverge.com/2015/9/23/9381509/ netflix-hooked-tv-episode-analysis.

CHAPTER 12: GAMIFICATION

293 **Volkswagen was releasing:** Website for DDB's Fun Theory campaign: www.thefunthe ory.com; Cannes Awards announcement: www.prnewswire.com/news-releases/ddbs-fun -theory-for-volkswagen-takes-home-cannes-cyber-grand-prix-97156119.html; video of the "Piano Stairs" experiment: Rolighetsteorin, "Piano Stairs: TheFunTheory.com," YouTube, October 7, 2009, www.youtube.com/watch?v=2lXh2n0aPyw.

295 **A broad survey:** Obesity data from the World Obesity Federation: www.worldobesity.org/ resources/obesity-data-repository; Kaare Christensen, Gabriele Doblhammer, Roland Rau, and James W. Vaupel, "Ageing Populations: The Challenges Ahead," *Lancet* 374, no. 9696 (October 2009): 1196–1208; John Bound, Michael Lovenheim, and Sarah Turner, "Why Have College Completion Rates Declined? An Analysis of Changing Student Preparation and Collegiate Resources," *American Economic Journal: Applied Economics* 2, no. 3 (July 2010): 129–57; Jeffrey Brainard and Andrea Fuller, "Graduation Rates Fall at One-Third of 4-Year Colleges," *Chronicle of Higher Education*, December 5, 2010, chronicle.com/article/Graduation -Rates-Fall-at/125614; World Bank savings data: data.worldbank.org/indicator/NY.GNS .ICTR.ZS; OECD savings data: data.oecd.org/hha/household-savings-forecast.htm; World Giving Index from the Charities Aid Foundation: www.cafonline.org/about-us/publications; report by National Center for Public Policy and Higher Education suggesting that the income of workforce expected to decline: www.highereducation.org/reports/pa_decline.

296 **A computer programmer named John:** On Breen and FreeRice: Michele Kelemen, "Net Game Boosts Vocabulary, Fights Hunger," NPR, December 17, 2007, www.npr.org/ templates/story/story.php?storyId=17307572.

298 **What DDB did:** Background on gamification and examples: Kevin Werbach and Dan Hunter, *For the Win: How Game Thinking Can Revolutionize Your Business* (Philadelphia,

PA: Wharton Digital Press, 2012), 168–72; Nick Pelling explains the history of the term: Nick Pelling, "The (Short) Prehistory of 'Gamificiation' . . . ," Funding Startups (& other impossibilities), Nanodome, April 9, 2011, nanodome.wordpress.com/2011/08/09/the-short -prehistory-of-gamification/; Dave McGinn, "Can a Couple of Reformed Gamers Make You Addicted to Exercise?" *Globe and Mail,* published November 13, 2011, last updated September 6, 2012, www.theglobeandmail.com/life/health-and-fitness/fitness/can-a -couple-of-reformed-gamers-make-you-addicted-to-exercise/article4250755/; Fox Van Allen, "Sonicare Toothbrush App Proves Too Addicting for Kids," Techlicious, September 16, 2015, www.techlicious.com/blog/philips-sonicare-for-kids-electric-toothbrush -app-sparkly/; Kate Kaye, "Internet of Toothbrushes: Sonicare Pipes Data Back to Philips," AdvertisingAge, September 14, 2015, http://adage.com/article/datadriven-marketing/ philips-connects-sonicare-kids-game-data-insights/300316.

302 **Q2L was the:** On Q2L, cognitive miserliness, and gamified education: Institute of Play, "Mission Pack: Dr. Smallz: Can You Save a Dying Patient's Life?" 2014, www.instituteof play.org/wp-content/uploads/2014/08/IOP_DR_SMALLZ_MISSION_PACK_v2.pdf; statistics on Q2L: Quest to Learn, "Research: Quest Learning Model Linked to Significant Learning Gains," www.q2l.org/about/research; Rochester Institute of Technology, Just Press Play, RIT Interactive Games & Media, play.rit.edu/About; Traci Sitzmann, "A Meta-analytic Examination of the Instructional Effectiveness of Computer-Based Simulation Games," *Personnel Psychology* 64, (May 2011): 489–528; Susan T. Fiske and Shelley E. Taylor, *Social Cognition Second Edition* (New York: McGraw-Hill, 1991); Dean Takahashi, "Study Says Playing Videos Games Can Help You Do Your Job Better," *New York Times*, December 1, 2010, www.nytimes.com/external/venturebeat/2010/12/01/01venturebeat -study-says-playing-videos-games-can-help-you-76563.html.

307 **After Rodney Smith:** Yagana Shah, "Story of a 93-Year-Old and 2 Lawn Mowers Will Melt Your Heart," Huffington Post, April 28, 2016, www.huffingtonpost.com/entry/story -of-a-93-year-old-and-2-lawn-mowers-will-melt-your-heart_us_572261aae4b0b49df6 aab03d; more on the badge T-shirt system at: Facebook, Raising Men Lawn Care Services Michigan, post, May 21, 2016, www.facebook.com/282676205411413/photos/a.28268973207 6727.1073741828.282676205411413/282689718743395/.

309 **The same properties:** Emily A. Holmes, Ella L. James, Thomas Coode-Bate, and Catherine Deeprose, "Can Playing the Computer Game 'Tetris' Reduce the Build-Up of Flashbacks for Trauma? A Proposal from Cognitive Science" *Plos One* 4, January 7, 2009, DOI: 10.1371/journal.pone.0004153; "Post-Traumatic Stress Disorder (PTSD): The Management of PTSD in Adults and Children in Primary and Secondary Care," London National Institute for Health and Clinical Excellence, 2005, CG026; J. A. Anguera and others, "Video Game Training Enhances Cognitive Control in Older Adults," *Nature* 501 (September 2013): 97–101; "Game Over? Federal Trade Commission Calls Brain-Training Claims Inflated," January 8, 2016, ALZforum, www.alzforum.org/news/community-news/game-over-federal-trade-commission-calls-brain-training-claims-inflated; but note this statement from detractors: Stanford Center on Longevity and the Max Planck Institute for Human Development, "A Consensus on the Brain Training Industry from the Scientific Community," October 20, 2014, longevity3.stanford.edu/blog/2014/10/15/the-consensus-on-the-brain-training-industry-from-the-scientific-community; a classic paper that explains why gamification may rob people of the intrinsic drive to behave in ways that benefit them: Uri Gneezy and Aldo Rustichini, "A Fine Is a Price," *Journal of Legal Studies* 29 (January 2000): 1–18.

313 **Bogost demonstrated the:** On Ian Bogost and Cow Clicker: The game's site: cowclicker .com; Bogost's own description of the game: bogost.com/writing/blog/cow_clicker_1/; see

also: Jason Tanz, "The Curse of Cow Clicker: How a Cheeky Satire Became a Hit Game," *Wired*, December 20, 2011, www.wired.com/2011/12/ff_cowclicker/all/1; interview with Bogost: NPR, "Cow Clicker Founder: If You Can't Ruin It, Destroy It," November 18, 2011, www.npr.org/2011/11/18/142518949/cow-clicker-founder-if-you-cant-ruin-it-destroy-it.

EPILOGUE

318 **This is known as the:** Oliver Burkeman, "This Column Will Change Your Life: The End-of-History Illusion," *Guardian*, January 19, 2013, www.theguardian.com/lifeandstyle/2013/jan/19/change-your-life-end-history; Jordi Quoidbach, Daniel T. Gilbert, and Timothy D. Wilson, "The End of History Illusion," *Science* 339, no. 6115 (January 2013) 96–98.

Index